Abdellah Ajji, Ebrahim Jalali Dil, Amir Saffar, Zahra Kanani Aghkand
Heat Sealing in Packaging

Also of Interest

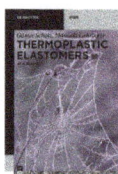

Thermoplastic Elastomers.
At a Glance
Scholz, Gehringer, 2021
ISBN 978-3-11-073983-1, e-ISBN (PDF) 978-3-11-073984-8

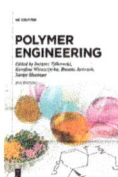

Polymer Engineering
Tylkowski, Wieszczycka, Jastrząb, Montane (Eds.), 2022
ISBN 978-3-11-073844-5, e-ISBN 978-3-11-073382-2

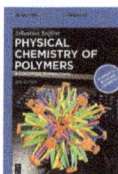

Physical Chemistry of Polymers.
A Conceptual Introduction
Seiffert, 2023
ISBN 978-3-11-071327-5, e-ISBN (PDF) 978-3-11-071326-8

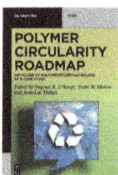

Polymer Circularity Roadmap.
Recycling of poly(methyl Methacrylate) as a Case Study
D'Hooge, Marien, Dubois, 2023
ISBN 978-3-11-071649-8

Plastics in the Circular Economy
Voet, Jager, Folkersma, 2021
ISBN 978-3-11-066675-5, e-ISBN (PDF) 978-3-11-066676-2

Abdellah Ajji, Ebrahim Jalali Dil, Amir Saffar,
Zahra Kanani Aghkand

Heat Sealing
in Packaging

——

Materials and Process Considerations

DE GRUYTER

Authors
Abdellah Ajji
Polytechnique Montreal
Chemical Engineering Department
C.P. 6079 Succursale Centre Ville
Montreal
Canada

Ebrahim Jalali Dil
PolyExpert Inc.
850 Munck Avenue
Laval H7S 1B1
Canada

Amir Saffar
ProAmpac
Boulevard des Entreprises
Terrebonne J6Y 1V2
Canada

Zahra Kanani Aghkand
Polytechnique Montreal
Chemical Engineering Department
C.P. 6079 Succursale Centre Ville
Montreal
Canada

ISBN 978-1-5015-2458-5
e-ISBN (PDF) 978-1-5015-2459-2
e-ISBN (EPUB) 978-1-5015-1616-0

Library of Congress Control Number: 2023931404

Bibliographic information published by the Deutsche Nationalbibliothek
The Deutsche Nationalbibliothek lists this publication in the Deutsche Nationalbibliografie;
detailed bibliographic data are available on the Internet at http://dnb.dnb.de.

© 2023 Walter de Gruyter GmbH, Berlin/Boston
Cover image: Amir Saffar
Typesetting: Integra Software Services Pvt. Ltd.
Printing and binding: CPI books GmbH, Leck

www.degruyter.com

Preface

The polymer packaging market reached $348 billion in 2020 and is growing at about 3–5% annually, especially in applications in the food industry. One of the main functions of a package is to contain and protect the (food) product during shipping and handling, delivering the product in the best condition intended for its use. Hence, an efficient sealing of the package is necessary in order to achieve this. In fact, sealing has such an important role in a package that if the sealing fails, the package fails and as a result, sealing of polymer films and packages has become the main concern in the design and development of polymer packages. We have been working on flexible packaging over the last 15 years, and heat sealing was and is still among the important topics of our research. Many of the results presented in this book originate from our previous works.

This book is intended to provide a comprehensive review that covers many aspects of heat sealing, from basic principles to materials and process. The main goal is to provide a sufficiently detailed and organized review of the heat-sealing process to obtain a deep insight into the involved molecular mechanisms as well as the effects of material and processing parameters on heat sealing. Readers wishing more details about any specific aspect discussed here can pursue further studies in the provided list of references in each chapter.

The molecular mechanisms involved in each step of heat sealing are discussed in detail in Chapter 2. Chapter 3 summarizes the sealing test methods used in the industry to evaluate the seal performance. Chapter 4 presents different materials commonly used as sealant layers in polymer packaging based on their seal performance. Chapter 5 reviews the literature on the effects of processing conditions as well as molecular characteristics of sealants on the seal and hot tack performance. Chapter 6 discusses the modeling and simulation of a heat-sealing process as a tool to reduce the required cost, energy, and time during sealants design for polymer packages.

Considering the much wider range of properties that polymer blends can offer compared to single-phase matrices, particularly in easy peel and peelable seals, sealants based on polymer blends have recently received an increasing attention. In Chapter 7, first, the concepts regarding microstructure–property relation in a polymer blend system are described, and then the seal performance of polymer blend sealants is discussed.

The environmental concerns regarding petroleum-based and non-biodegradable polymers that are conventionally used in polymer packages led to a significant interest in replacing these materials with bioplastic (biobased and/or biodegradable/compostable) sealants. Chapter 8 introduces bioplastic materials used as sealants in polymer packaging and reviews their seal behavior. Finally, Chapter 9 is dedicated to case studies in designing sealant layers in some applications to present guidelines for practical sealant film design.

https://doi.org/10.1515/9781501524592-202

Contents

Preface —— V

Chapter 1
Introduction —— 1
1.1   History of polymers in packaging —— 1
1.2   Plastic joining methods —— 2
1.2.1   Ultrasonic sealing —— 2
1.2.2   Induction sealing —— 4
1.2.3   Heat bar sealing —— 4
1.2.3.1   VFFS packaging machine —— 5
1.2.3.2   HFFS packaging machine —— 7
1.2.4   Impulse sealing —— 9
1.3   Organization of the book —— 10
            References —— 10

Chapter 2
Molecular mechanism of heat sealing —— 13
2.1   Heat sealing from microscopic viewpoint —— 13
2.2   Melting of polymer materials —— 15
2.3   Surface rearrangement —— 17
2.4   Molecular interdiffusion across the interface —— 18
2.4.1   Polymer interdiffusion dynamics —— 18
2.4.2   Interdiffusion at polymer–polymer interfaces —— 22
2.5   Cooling and crystallization —— 23
            References —— 27

Chapter 3
Seal quality and performance evaluation methods —— 31
3.1   Seal quality tests —— 31
3.1.1   Visual inspection —— 33
3.1.2   Flat bar test —— 33
3.1.3   Gross leak or bubble test —— 34
3.1.4   Pressure decay leak test —— 35
3.1.5   Dye penetration test —— 36
3.1.6   Pressure-assisted dye penetration test —— 37
3.1.7   Airborne ultrasound —— 37
3.2   Seal performance tests —— 38
3.2.1   Hot tack test —— 39

3.2.2   Seal strength measurement —— **42**
3.2.3   Internal pressurization failure resistance —— **44**
             References —— **46**

Chapter 4
Sealant layer materials —— 49
4.1   Polyethylene —— **49**
4.1.1   High-density polyethylene (HDPE) —— **50**
4.1.2   Low-density polyethylene (LDPE) —— **50**
4.1.3   Linear low-density polyethylene (LLDPE) —— **51**
4.1.4   Metallocene polyethylene (mPE) —— **53**
4.1.5   Plastomers —— **54**
4.2   Polypropylene (PP) —— **54**
4.3   Ethylene vinyl acetate (EVA) —— **56**
4.4   Acid copolymers —— **58**
4.5   Ionomers —— **59**
4.6   Polyethylene terephthalate (PET) —— **61**
             References —— **61**

Chapter 5
Effect of processing and material properties on seal performance —— 63
5.1   Sealing temperature —— **63**
5.2   Dwell time —— **66**
5.3   Sealing pressure —— **67**
5.4   Effect of material characteristics —— **69**
5.4.1   Effect of crystallinity —— **69**
5.4.2   Effect of molecular architecture —— **72**
5.4.3   Chain branching —— **72**
5.4.4   Monomer sequence —— **74**
             References —— **74**

Chapter 6
Modeling of heat sealing process —— 77
6.1   Modeling of the interface temperature —— **77**
6.1.1   Heat transfer in contact areas —— **79**
6.1.2   Heat transfer within the film —— **82**
6.2   Modeling of squeeze-out flow —— **85**
6.3   Modeling seal strength development —— **93**
             References —— **96**

Chapter 7
Multicomponent sealant films —— 101
7.1   Thermodynamics of polymer blends —— 101
7.2   Morphology of polymer blends —— 105
7.3   Surface morphology of immiscible polymer blend films —— 107
7.4   Sealants based on immiscible polymer blends —— 109
7.5   Polymer nancomoposites —— 112
             References —— 115

Chapter 8
Bioplastic sealants —— 119
8.1   Bio-based polyethylene terephthalate (bio-PET) —— 120
8.2   Bio-based Polyethylene (PE) and Ethylene Vinyl Acetate (EVA) —— 120
8.3   Poly(lactide) —— 121
8.4   Polycaprolactone —— 122
8.5   Aliphatic polyesters and copolyesters —— 122
8.6   Aromatic copolyesters —— 124
8.7   Polyhydroxyalkanoates (PHA) —— 125
             References —— 126

Chapter 9
Case studies —— 129
9.1   Design of a peelable sealant for cereal packaging —— 129
9.1.1   Solution —— 129
9.2   Sealant for liquid packaging —— 131
9.2.1   Solution —— 131
9.3   Film for heavy-duty shipping sack (HDSS) —— 131
9.3.1   Solution —— 132
9.4   Peelable film for over-the-mountain packaging —— 132
9.4.1   Solution —— 133
9.5   Sealant film with caulkability for spouted pouch application —— 134
9.5.1   Solution —— 135
9.6   Peelable barrier film for dried seeds packaging —— 136
9.6.1   Solution —— 136
9.7   Oxygen barrier sealant for cheese packaging —— 138
9.7.1   Solution —— 138
9.8   Film for frozen vegetables —— 139
9.8.1   Solution —— 139
             References —— 140

Index —— 141

Chapter 1
Introduction

1.1 History of polymers in packaging

The first flexible transparent polymeric film called cellophane was produced from regenerated cellulose by the Swiss chemist Jacques E. Brandenberger in the 1900s [1]. Cellophane was later commercialized by DuPont for food packaging applications; however, due to its sensitivity to moisture, it could not be used widely in the food industry. The first moisture barrier transparent polymer film was prepared by coating cellophane with nitrocellulose [2]. The first transparent oxygen and moisture barrier polymer film was then introduced by coating polyvinylidene chloride (PVDC) on cellophane [3]. This allowed cellophane to be introduced as a competitor for traditional metal barrier films in the food industry with the advantage of transparency and lightweight. Eventually, polyvinyl chloride (PVC) and polypropylene replaced cellophane in the packaging industry in the 1960s due to the poor performance of cellophane at low temperatures, its limited shelf-life, and cost issues. In 1973, the Food and Drug Administration banned the usage of PVC in beverage bottles due to migration of residual vinyl chloride monomers [4], and the application of PVC in food packaging decreased dramatically. The interesting properties of polyethylene (PE), including its chemical resistance, odorless, toughness, good sealability, high moisture barrier, and low cost, made it a strong candidate for food packaging applications. Since then, many other polymers have been widely used in the polymer packaging market, including poly(ethylene terephthalate), ethylene vinyl acetate, and polyamide. In addition, different types of coatings including thin PVDC or metal/mineral oxide coating (with commercial names such as ALO_x and SiO_x) allowed combining very high barrier properties and transparency in plastic packaging without using opaque aluminum foil layer. The growing needs, especially in the food industry [5], have significantly increased the market value of polymer packaging that finally reached $348 billion in 2020 [6]. This indicates the significant role of polymer packaging in the future of the food industry.

The main functions of a package can be listed as containment of the product, protecting the product during shipping and handling, delivering the product in the best condition intended for its use, and finally, brand advertisement [5]. An efficient sealing of a package is necessary in order to achieve the first three roles of the package. In fact, sealing has such an important role in a package that if the sealing fails, the package fails. As a result, sealing of polymer films and packages has become a main concern in their design and development. The importance of sealability is such important that the lack of sealing properties delayed the use of certain polymers in the past. For example, despite interesting moisture barrier properties of nitrocellulose-coated cellophane, this film was used widely in packages only after modifications that

https://doi.org/10.1515/9781501524592-001

allowed it to be heat-sealable [3]. Different techniques have been proposed for sealing polymer packages [7] that will be briefly reviewed below.

1.2 Plastic joining methods

Different techniques have been used to join plastic components ranging from mechanical joining using screws, to adhesive bonding and welding techniques. In the welding methods, polymeric surfaces are melted, and surfaces are joined by polymer chain diffusion across the interface. Welding methods can be categorized into external heating or internal heating methods depending on the method of applying heat. Figure 1.1 shows the most common welding techniques based on their heating source.

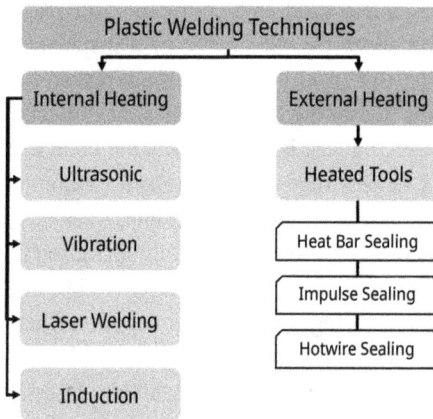

Figure 1.1: Different welding techniques based on applying heat methods.

In the following sections, we review ultrasonic sealing, induction sealing, heat bar sealing, and impulse sealing as they are mostly used in sealing plastic packaging.

1.2.1 Ultrasonic sealing

Ultrasonic sealing was initially developed by Robert Soloff in the 1960s. In this technique, two sides of the seal area are pushed together between an ultrasonic horn and an anvil. Then the horn vibrates at a high frequency in the range of 20–40 kHz and amplitude in the range of 10–30 μm [9]. The main components of an ultrasonic sealing machine are a transducer, a booster, and a horn (known as sonotrode). The transducer is a piezoelectric material that converts the high-frequency electric signal to high-frequency mechanical vibration. This vibration is amplified by the booster and transferred to the sonotrode. The vibration results in molecular movement and heat generation due to intermolecular friction. Energy directors with different shapes are used to concentrate the vibration to the desired sealing area. Consequently, the temperature in the seal area is raised, which melts the sealant material. Under the applied pressure by the ultrasonic horn, molecular diffusion across the interface occurs and seals the two sides of the seal area. The sealed area is cooled down under force between horn and anvil after stopping the vibration, and finally, the horn will be opened. Figure 1.2 schematically shows a typical ultrasonic sealing process.

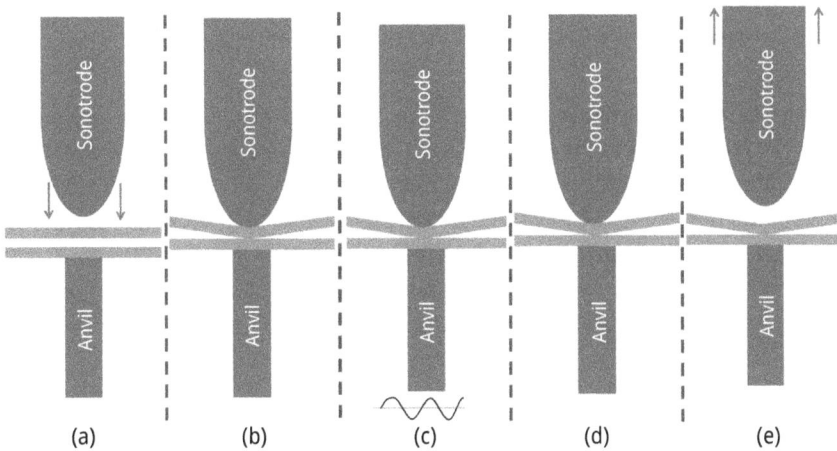

Figure 1.2: Ultrasonic sealing steps: (a) approaching, (b) applying force, (c) starting of ultrasonic vibration, (d) end of ultrasonic vibration and beginning of cooling under force, and (e) removing of the force and opening of sonotrode.

Ultrasonic sealing has been used in a wide range of applications from aerospace to medical and food packaging. Ultrasound sealing is known to provide a great seal in the presence of contamination and is an effective sealing approach for thick samples. As in ultrasonic sealing, the heat does not transfer from outside to the inside of the product and is generated directly at the sealing interface, it allows a much faster sealing process especially in sealing of thick samples such as multiwall paper bag products. The lack of the need for heat transfer from outside of the film also allows safe sealing for temperature-sensitive materials such as medications. Some studies also showed that ultrasound sealing needs less energy consumption compared to heat sealing [9].

1.2.2 Induction sealing

This type of sealing is also known as cap sealing and is very common in the sealing of rigid containers' lidding. In this process, the lidding film is made of a structure that contains four different layers as shown in Figure 1.3.

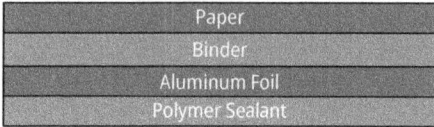

| Paper |
| Binder |
| Aluminum Foil |
| Polymer Sealant |

Figure 1.3: A common structure for lidding is used in the induction sealing process.

A schematic of an induction sealing process is shown in Figure 1.4. The lidding is placed on the container and the cap is closed. Then the container passes through an induction chamber or induction zone using a conveyor belt. An electromagnetic field is applied in the induction chamber which interacts with the foil layer and generates heat. The generated heat melts the sealant layer and results in the adhesion of the lidding to the container. In addition, the generated heat melted the binder which is then infiltrated into the paper and resulted in releasing of aluminum from the backing upon cooling.

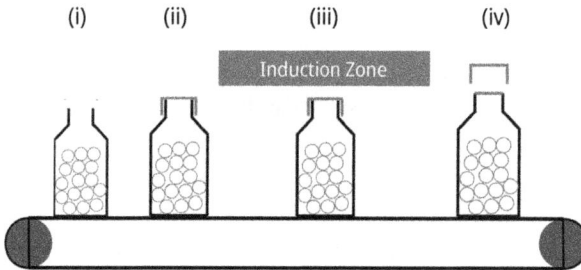

Figure 1.4: Different steps in the induction sealing process: (i) filled container, (ii) closing of the cap with lidding, (iii) sealing of the cap to the container in the induction zone, and (iv) sealed container after the opening of the cap.

1.2.3 Heat bar sealing

Heat bar sealing is the most common type of sealing in plastic packaging due to its ease of use and low maintenance cost. Figure 1.5 schematically shows a heat bar sealing process. In this method, two sides of the seal area are pushed together between two jaws (bars) that both or only one of them is heated. To seal both sides together, the jaws apply a certain pressure, known as the sealing pressure, for a certain period of time, known as the dwell time or the sealing time. The first sealing machine with electrically heated jaws was invented in 1942 for the sealing of cellophane bags [3].

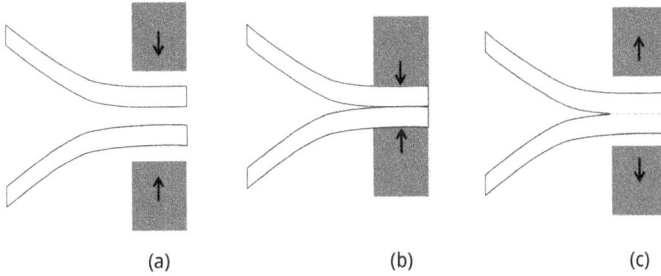

(a)   (b)   (c)

Figure 1.5: Schematics of the heat bar sealing process: (a) approaching the heated jaws, (b) applying heat and pressure for the duration of dwell time to the seal area, and (c) opening of the jaws.

The main processing parameters in the heat bar sealing process are sealing temperature, sealing pressure, and dwell time. In addition, the pattern of the surface of jaws can affect seal properties. For example, Selke et al. [10] and Theller [11] showed that using serrated jaws with patterns could improve seal performance by increasing the intimate contact between polymeric films especially when the thickness of the films is not uniform.

Heat bar sealing can be found in very different packaging machines. Vertical form fill and seal (VFFS), horizontal form fill and seal (HFFS), and pouch-making machines are the most common packaging machines that use heat bar sealing. A brief explanation of VFFS and HFFS processes will be given in the next sections. The same concepts can be applied for pouch-making machines. In addition to these machines, heat bar sealing is also used in different bag-making machines, pouch-making machines and tray-sealing machines.

1.2.3.1 VFFS packaging machine

Figure 1.6 schematically shows how a VFFS machine is used to form, fill, and seal a package. In this machine, the film is fed in the form of a sheet and passes over a metal collar to form a tube. The formed tube passes over a metal tube that is known as a forming tube where a vertical sealer seals two edges of the film together to form a complete plastic tube around the forming tube.

The films can be sealed in a vertical sealer by sealing inside to inside (known as the fin seal) or outside to inside (known as the lap seal). These two types of vertical sealing are schematically shown in Figure 1.7.

Sealing of the vertical seal is commonly done by the heated bar, but other types such as hot air are also used in special cases. After passing through the vertical sealer,

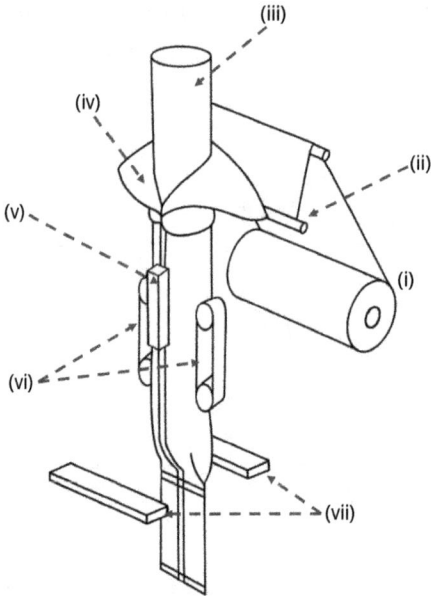

Figure 1.6: Different parts of a VFFS machine: (i) plastic film roll, (ii) dancing arm, (iii) forming tube, (iv) forming collar (shoulder), (v) vertical sealer, (vi) transport belts, and (vii) heated sealing bars.

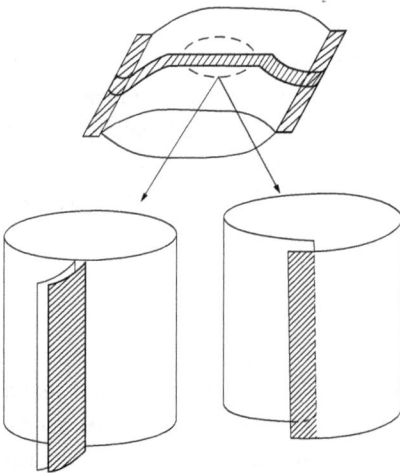

Figure 1.7: Different sealing methods for vertical seal in VFFS machine: (left) fin seal and (right) lap seal. The hatched area shows the sealed area.

the film reaches the end of the forming tube where the end of the plastic tube is sealed by a pair of heated jaws to create the bottom seal of the bag. The film then is transported downward to reach the desired bag height. The top of the bag is then sealed, and a knife that is inside of the jaws cuts the bag off the plastic tube. In VFFS machines, the film transport can occur by two means: (i) the jaws grip and pull down the film, known as the jaw draw-off principle; (ii) a belt is installed on the sides of the forming tube and pulls down the film, known as belt pick-up principle.

In the first transport form, when the jaws are closed to seal the bottom of the plastic tube, they also have a simultaneous downward movement which results in pulling down the film to the desired bag height. Therefore, in this form of transport, the friction between the inside layer of the films and the forming tube is very critical. The machines that use a jaw draw-off mechanism cannot produce bags with block bottom.

In the belt pick-up mechanism, the film is pushed down due to friction between belts located on the sides of the forming tube. Therefore, in this form of film transport, the friction between the inside of the films and the forming tube surface as well as the friction between the belt and the outer surface of the film becomes important. This has led to the development of the vacuum transport belts in which the film is sucked toward the belt to reduce the friction between the film and the forming tube and improve the friction between the belt and the film. One interesting advantage of the belt principle over jaw drawdown is that it can be used for making any type of bag.

VFFS machines are mostly intermittent machines with a pause in their production. In the jaw drawdown machines, this time is needed for the jaws to open and return to their initial sealing position to seal the top of the bag. In machines with the belt pickup mechanism, this pause is needed for the jaw to open and allow downward film transport equal to the length of the bag. To create such an intermittent motion, most machines are equipped with what is known as a dancer arm mechanism to translate the continuous unwinding of the plastic roll into the intermittent motion. In intermittent VFFS processes, the vertical sealing is also intermittent, which means the vertical sealer touches the film on the forming tube at the same time as the bottom sealer seals the bottom of the tube and retracts back during the transportation step. This motion allows film transport and prevents overheating of the vertical sealing part during the transport phase of the process. The intermittent process reduces productivity and increases production costs. By reducing the size of the package, the time-lapse between steps is reduced, and in small snack size packaging, the process is almost continuous. In some machines, the jaw drawdown and belt pickup mechanisms are combined to provide continuous film motion. A new patented technology called twin-jaw continuous motion VFFS was introduced, in which four heated jaws are mounted on two circular shafts. The jaws rotate in a synchronized manner to provide continuous film motion without a delay for sealing. This reduces considerably the delay related to jaw motion and could reach 300 bags per minute packaging speed.

1.2.3.2 HFFS packaging machine

Figure 1.8 and 1.9 schematically show the process flow diagram of the two most common types of HFFS packaging machines: flow wrapper and HFFS pouch machines.

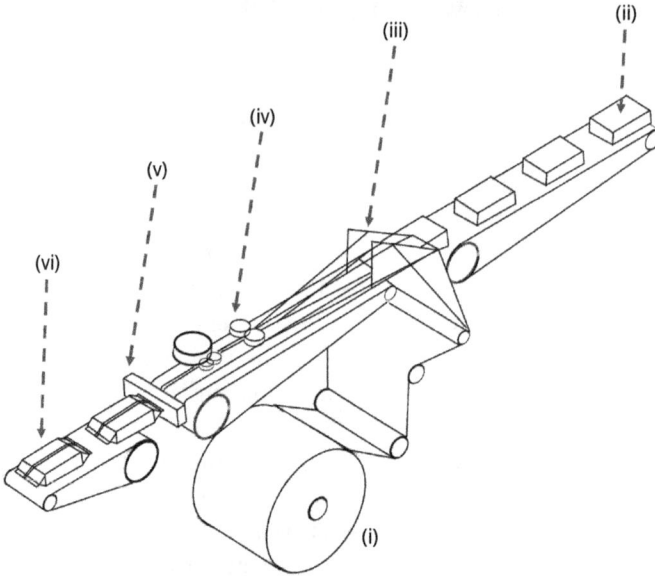

Figure 1.8: Flow wrapper: (i) plastic roll, (ii) product, (iii) bag/forming, (iv) side/vertical sealers, (v) bottom/horizontal sealer, and (vi) final packaged product.

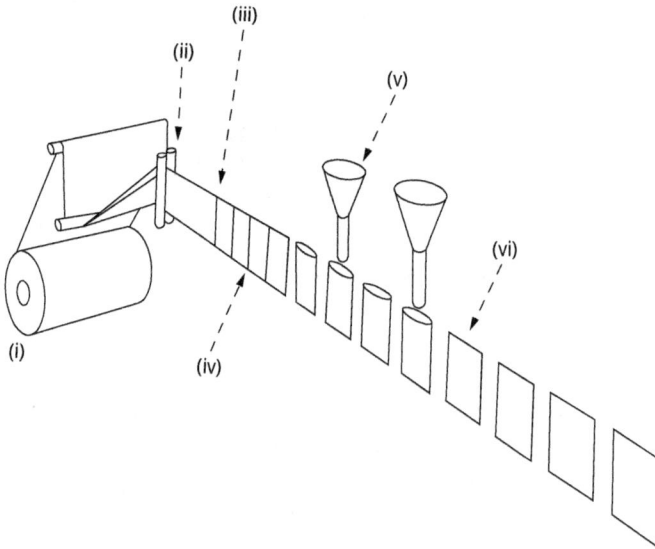

Figure 1.9: HFFS pouch machine: (i) film roll, (ii) film folding and gusset forming, (iii) vertical side sealing, (iv) horizontal sealing, (v) filling and top sealing, and (vi) cutting and packaging.

In the flow wrapper machine, the film passes over a forming box to form a tube-like structure where the edges pass between rolls beneath the product's conveyor belt. The edges are sealed by the center sealer beneath the platform. The product is fed from the center of the forming box into the formed plastic tube. The package end is sealed using the end sealer, which is usually a rotating heated seal bar.

In pouch-making HFFS machine, the film is folded and, if needed, gusseted. Then a vertical heat bar seal seals the side sealing of the package and a horizontal sealing seals the bottom (if needed). The packaging is then filled from the top and then a horizontal heat bar seals the top of the package. Finally, the pouch is cut and packaged. The HFFS pouch machine can produce pouches with or without zipper. Using HFFS pouch machine allows combining pouch-making and filling step in one step, which reduces the time and associated costs.

1.2.4 Impulse sealing

Impulse sealing is a very cheap sealing machine that can be found in kitchens and small businesses. The impulse sealing machine consists of two jaws, where one jaw is heated using an electric wire passing beneath a Teflon cover and the other jaw is used to apply the sealing pressure (Figure 1.10).

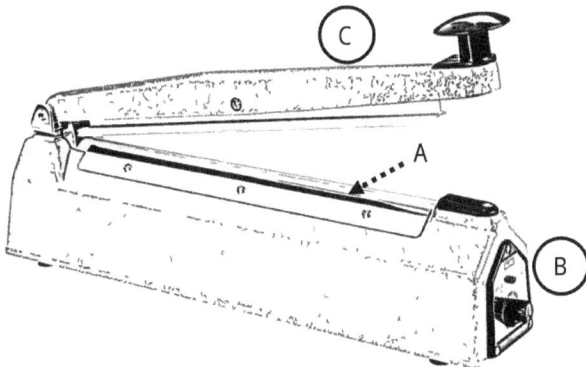

Figure 1.10: Manual impulse heat sealer: (A) Teflon-coated sealing wire; (B) dwell time controller; and (C) pressure applying handle.

When the films are placed between moving and fixed jaws, and after the closing of the upper jaw, the current in the sealing wire is switched on automatically. The heating wire is usually made of Nichrome and the heat is generated in the wire due to its resistance against electrical current. The current remains for the desired time selected by the dwell time controller or until the jaw is opened by the user depending on the machine design. In this type of basic impulse heat sealer, control on temperature and dwell time is very poor. On the other hand, as the heating is only applied once the jaw is closed, this type of heat sealing reduces power consumption compared to constantly heated jaws in heat bar sealing. In addition, impulse sealers do not require a warm-up time, which allows for quick start-up.

1.3 Organization of the book

Among different sealing methods in plastic packaging, heat sealing, especially heat bar sealing, is the most common method used in the polymer packaging industry due to its convenience, high speed, and low operation cost [8]. Despite its importance in the packaging industry, a comprehensive review that covers all aspects of heat sealing is lacking in the literature. The main goal of this book is to provide a detailed and organized review of the heat-sealing process to obtain a deep insight into the involved molecular mechanisms as well as the effects of material and processing parameters on heat sealing. In all parts of the book, these concepts are discussed by considering their implications in real-life and industrial applications to establish a relation between knowledge and application.

The molecular mechanisms involved in each step of heat sealing are discussed in detail in Chapter 2. Chapter 3 summarizes the test methods used in the industry to evaluate the seal performance. Chapter 4 presents different materials commonly used as sealant layers in polymer packaging based on their seal performance. Chapter 5 reviews the literature on the effects of processing conditions as well as molecular characteristics of sealants on the seal and hot tack performance. Chapter 6 discusses the modeling and simulation of a heat-sealing process as a tool to reduce the required cost, energy, and time during sealant design for polymer packages. Multicomponent sealant films are discussed in Chapter 7, as new seal material generation allows widening seal performance spectrum. Due to the environmental concern of plastic waste, bioplastic sealant materials have been used to produce environment-friendly packaging. This type of sealant materials will be discussed in Chapter 8. Chapter 9 presents different case studies and troubleshooting to allow readers to better understand the implications of the discussed materials.

References

[1]   Brandenberger, J.E., Tube en cellophane destiné à contenir des matières grasses FR451485A, 1913.
[2]   Hale, C.W. and P.K. Edwin, Moisture proofing composition, US1826696A, 1931.
[3]   Kane, W.P., *Cellophane Coating Compositions Comprising Vinylidene Chloride Copolymer, Candelilla Wax and Stearate Salt*. 1968, Google Patents.
[4]   De La Cruz, P.L., *PVC: Packaging issues and regulations*. Journal of Vinyl Technology, 1988. **10**(3): p. 117–120.

[5]   Robertson, G.L., *Food Packaging: Principles and Practice*. 2013, CRC Press.

[6]   *Plastic Packaging (Rigid Plastic Packaging and Flexible Plastic Packaging) Market for Food & Beverages, Industrial, Household Products, Personal Care, Medical and Other Applications – Global Industry Perspective, Comprehensive Analysis, Size, Share, Growth, Segment, Trends and Forecast*. 2016, Zion Market Research. https://www.marketresearchstore.com/report/plastic-packaging-market-z47941.

[7]   Troughton, M.J., *Handbook of Plastics Joining: A Practical Guide*. 2008, William Andrew.

[8]   Morris, B.A., *Chapter 7: Heat Seal, in the Science and Technology of Flexible Packaging*. 2017, Elsevier, Cambridge.

[9]   Bach, S., et al., *Ultrasonic sealing of flexible packaging films–principle and characteristics of an alternative sealing method*. Packaging Technology and Science, 2012. **25**(4): p. 233–248.

[10] Selke, S.E.M., J.D. Culter, and R.J. Hernandez, *Plastics Packaging: Properties, Processing, Applications, and Regulations*, ed. 2nd. 2004: Hanser Gardner Publications.

[11] Theller, H.W., *Heat sealability of flexible web materials in hot-bar sealing applications*. Journal of Plastic Film and sheeting, 1989. 5: p. 66–93.

Chapter 2
Molecular mechanism of heat sealing

2.1 Heat sealing from microscopic viewpoint

In this chapter, the mechanisms involved in heat sealing of polymeric films are discussed in detail. The following steps can be considered in a common heat bar sealing process [1, 2]: (i) surface contact between two films; (ii) heat transfer from the jaw(s) toward the interface between films; (iii) surface melting and rearrangement; (iv) molecular interdiffusion across the interface; and finally (v) cooling and recrystallization after opening the jaws. Figure 2.1 shows a schematic of these steps from the microscopic viewpoint.

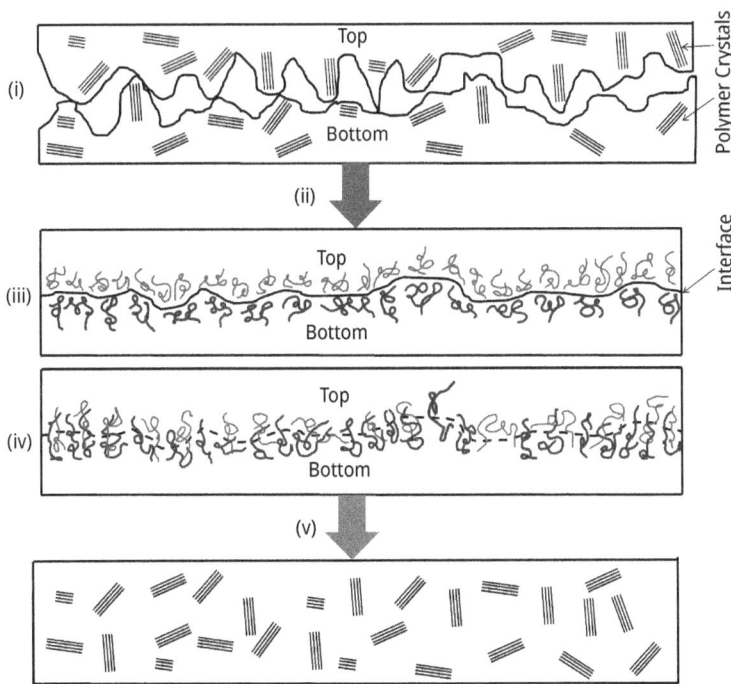

Figure 2.1: Different steps in a heat sealing process: (i) surface contact, (ii) heat transfer from the jaws, (iii) surface melting and rearrangement, (iv) molecular interdiffusion across the interface, and (v) cooling and recrystallization after opening the jaws.

In the first step, two sealing sides are pushed together by the heated jaws. However, as in the early stages of the process, the interface temperature (T_{int}) is below the melting temperature (T_m); therefore, the micro-roughness on the surfaces limits the contact between two films at the interface. Figure 2.2 shows an example of the microroughness

https://doi.org/10.1515/9781501524592-002

Figure 2.2: Microroughness profile on the surface of a polyethylene sealant film.

that exists on the surface of a polyethylene sealant film with an average square mean roughness (R_q) of 95 nm.

Reducing the roughness of the surfaces, reducing polymer hardness, and increasing the applied sealing pressure are known strategies to increase the contact area in this step [3]. However, due to the low interface temperature in this step, the molecular mobility at the interface is very limited, and chain diffusion across the interface does not occur. As a result, very poor seal strength is obtained if the heat sealing process is stopped at this stage. Interface temperature increases by increasing the dwell time due to the heat transfer from the heated jaws. Figure 2.3 schematically shows a typical interface temperature profile in a heat sealing process. The interface temperature increases rapidly at the

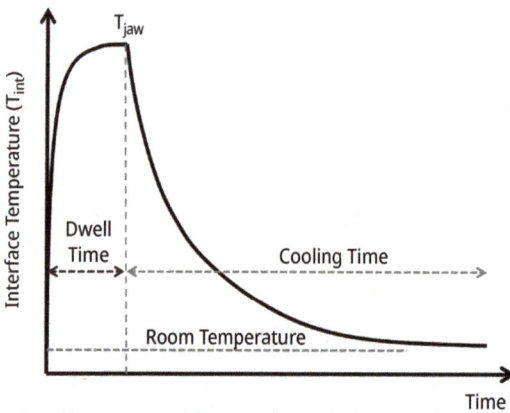

Figure 2.3: Typical interface temperature variation in a heat sealing process. T_{jaw} shows the jaw temperature.

early stages of the process. Heat transfer from the jaws depends on many parameters, including jaw interface with films, jaw temperature, dwell time, thermal conductivity of polymers, film thickness, and polymer crystallinity. The effect of these parameters on heat transfer in the heat sealing process will be discussed in Chapter 5.

When the interface temperature reaches the melting temperature of the sealant, a significant deformation in surface asperities occurs due to the applied pressure by jaws. This allows surface rearrangement and increases the contact area between the two films to its maximum. In addition, melting of polymer crystals increases significantly the molecular mobility of polymer chains and, after surface rearrangement and establishing surface contact, allows interdiffusion across the interface and formation of molecular entanglements. In the last step, after opening the jaws, temperature in the sealed area decreases due to the heat transfer to the surrounding medium. When T_{int} falls below the crystallization temperature (T_c) of the sealant, the polymer chains begin to form polymer crystals. The formed crystals act as physical cross-linking points and reduce the molecular mobility which increases the seal strength [1, 3].

As will be shown later, the hot tack strength is measured right after opening of the jaws when the sealant is still in the molten state; therefore, hot tack depends mainly on chain interdiffusion across the interface (step iv in Figure 2.1). On the other hand, seal strength is affected by both molecular interdiffusion and recrystallization after jaw opening as seal strength is measured when the sealant reaches room temperature. As mentioned previously, molecular interdiffusion begins only after melting and surface rearrangement; therefore, first, melting of polymer materials will be briefly discussed, and then surface rearrangement, interdiffusion, and finally, crystallization will be reviewed in the rest of this chapter.

2.2 Melting of polymer materials

From the thermodynamic point of view, the melting phenomenon can be explained by considering the Gibbs free energy of fusion:

$$\Delta G_f = \Delta H_f - T \Delta S_f \tag{2.1}$$

where ΔH_f and ΔS_f are molar enthalpy and entropy variations and T is the temperature. As ΔG becomes zero at the melting temperature of crystals, the melting temperature can be defined as

$$T_m = \frac{\Delta H_f}{\Delta S_f} \tag{2.2}$$

This equation simply shows how changing the molecular parameters such as M_w (which changes ΔS_f) or intermolecular interactions (which affects ΔH_f) can alter the melting temperature of crystals. At $T \geq T_m$, the ordered crystal structure breaks down and is transferred into a disordered amorphous phase, which leads to an increase in

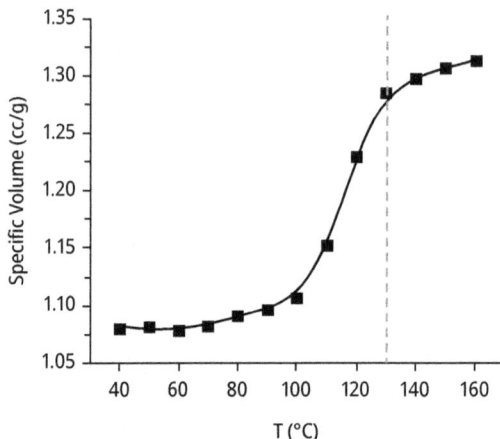

Figure 2.4: Variation of specific volume of a high-density polyethylene (HDPE) with temperature, and the dashed line shows the peak melting temperature determined from DSC (data from [5]).

the specific volume of the polymer. This causes a discontinuity in the variation of specific volume with temperature. As polymer crystals have a distribution of crystal size and crystal perfection due to their high molecular weight (Mw) and molecular weight distribution (MWD) [4], the variation of the specific volume–temperature curve of polymers is usually seen as a transition similar to the one shown in Figure 2.4 rather than a sharp discontinuity at the melting temperature.

The most common and practical method in determining the melting temperature of polymer materials is dynamic scanning calorimetry (or DSC) analysis. Melting is an

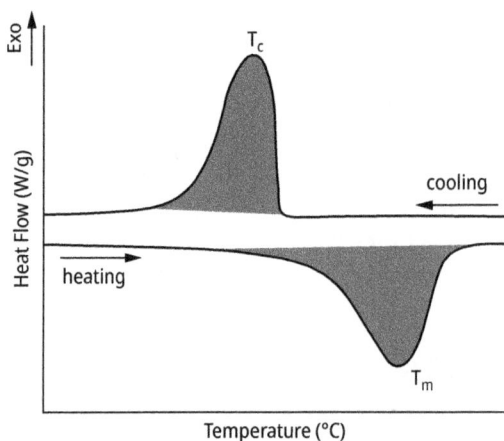

Figure 2.5: A typical DSC curve of a semicrystalline polymer material showing the peak melting temperature (T_m) and the crystallization temperature (T_c). The marked areas under peaks show the enthalpies of melting and crystallization.

endothermic transition; therefore, the DSC technique can be used to determine this transition from the absorbed heat of fusion. Figure 2.5 shows a typical DSC curve of a polymer material with the peak melting temperature (T_m). As mentioned previously, the distribution of crystal size and their different perfection level results in a broad melting peak in polymers. Therefore, molecular parameters such as Mw and MWD [6–8] as well as processing conditions such as cooling rate from the melt [9–12] and orientation [13, 14] influence the breadth of the melting peak in the DSC curve.

It should be noted that the melting temperature and heat of fusion reported from the DSC data are usually recorded at heating rates of 10 or 20 °C/min recommended in the ASTM standard [15]. However, polymer materials experience extremely higher heating rates in a heat sealing process. As an example, when the jaw temperature is set at 140 °C and for a dwell time of 0.5 s, polymer materials experience a heating rate of 230 °C/s or 13,800 °C/min. Previous studies showed that at heating rates lower than 100 °C/min, increasing the heating rate increased the melting temperature and the breadth of the melting peak of polymer materials [16]. Recent advances in DSC technology allowed studying melting of polymers at extremely high heating rates up to 6×10^6 °C/min [12, 17, 18]. It has been shown that exposing polymer materials to extremely high heating rates can alter their melting behavior in different ways [12, 18]. For instance, Schawe [19] showed that increasing the heating rate from 600 to 12,000 °C/min reduced the melting temperature of isotactic polypropylene (iPP) by almost 10 °C but further increase in the heating rate increased the melting temperature. On the other hand, Hellmuth and Wunderlich [20] found that increasing the heating rate from 10 to 1,000 °C/min did not have a considerable effect on the observed melting temperature of polyethylene. Minakov et al. [21] reported that increasing the heating rate from 60,000 to 216,000 °C/min increased the melting temperature of polyethylene terephthalate by 4 °C. They also studied superheating of isotactic polystyrene (PS) and found a 10 °C increase in its melting temperature in the abovementioned heating rates. As a general conclusion, it can be seen that despite different reported behaviors of polymers exposed to high heating rates, the previous studies showed that the variation of melting temperature was always less than 10 °C, which indicates that the effect of high heating rates on the melting point of polymers should not be significant.

2.3 Surface rearrangement

At temperatures above the melting point and under pressure applied by jaws, asperities at the surface of films disappear, and a complete surface contact between the two sides of the seal is established. In this step, the interface formation between two films occurs through wetting and spreading mechanisms. Spreading and wetting begin from the regions with the lowest surface roughness and then grow in the two-dimensional interface plane [22–24]. This is shown schematically in Figure 2.6. The rate of surface wetting increases by increasing the temperature, pressure, and chain mobility [22, 24–28].

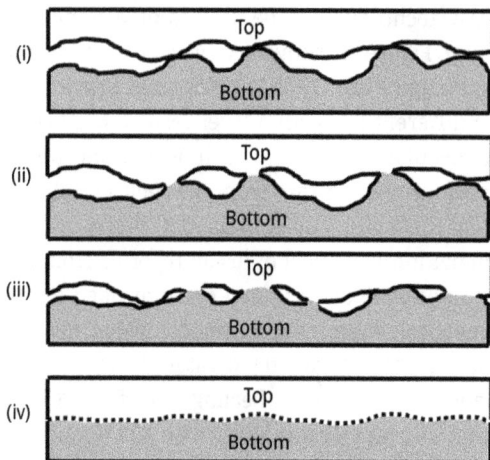

Figure 2.6: Schematic of wetting and spreading process in 2D plane of the interface between two seal sides.

The interdiffusion step begins only after establishing contact between two seal sides through surface rearrangement. This indicates the importance of surface rearrangement step in a heat sealing process. The surface rearrangement and wetting can be hindered by surface impurities resulting from processing or handling of the films.

2.4 Molecular interdiffusion across the interface

At temperatures above T_m of the sealant and after establishing surface contact through surface rearrangement, polymer chains at the interface will have the opportunity to diffuse across the interface. The interface healing and strength development are directly related to molecular interdiffusion across the interface [29]. Therefore, understanding the mechanism and kinetics of molecular diffusion in polymer melts is critical in analyzing the heat sealing process. For example, estimation of the minimum time required for interdiffusion and formation of entanglements across the interface allows minimizing dwell time and increasing the production rate.

2.4.1 Polymer interdiffusion dynamics

Different models have been proposed to explain polymer chain dynamics in the solution and the bulk. The first model that tried to explain polymer chain motion was presented by Rouse [30] in which a polymer chain is divided into submolecule segments. Each segment consists of polymer units (mers) and has enough length to be considered as a

Gaussian chain. This model was later reformulated by Bird et al. [31] in the form of successive repetition of beads and Hookean springs similar to the one shown in Figure 2.7.

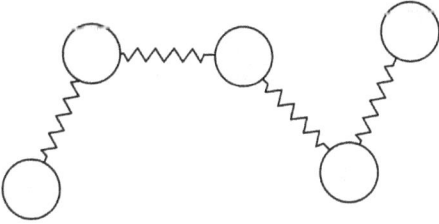

Figure 2.7: The schematic representation of the Rouse model including beads and Hookean springs.

In this model, it is assumed that the beads and springs are surrounded by a viscous medium which, in the case of polymer melts, represents the surrounding polymer chains. This model allows a significant internal degree of freedom and can explain chain orientation and extension. An important parameter in this model is the Rouse relaxation time which is defined as

$$\tau_R = \frac{6\,\eta\,M}{\pi^2 \rho R T} \qquad (2.3)$$

where ρ, M, R, η, and T are density, Mw, universal gas constant, viscosity, and temperature, respectively. The diffusion coefficient of a polymer chain (D_c) in this model is defined as $D_c = D/M$, where D is the diffusion coefficient of a statistical segment. Based on the Rouse model, the mean square displacement of a polymer chain, $<X^2>$, obeys two regimes:

(i) at $t < \tau_R$:

$$<X^2> = \left(\frac{12MD_c b^2}{\pi}\right)^{1/2} t^{1/2} \qquad (2.4)$$

(ii) at $t > \tau_R$:

$$<X^2> = 6D_c t \qquad (2.5)$$

In equation (2.4), b is the statistical chain segment length. Considering the definition of D_c, it can be seen that the movement of a polymer chain at $t < \tau_R$ is independent of the mass of the chain and depends on time by $t^{1/2}$. On the other hand, at $t > \tau_R$, the Rouse model predicts that the displacement of a chain is inversely related to its mass and depends linearly on time. In fact, based on the Rouse model, at $t > \tau_R$, the whole chain diffuses and, as a result, the chain movement is independent of the statistical segment selection.

As the Rouse model does not consider the effect of chain–chain interactions, it cannot predict the effect of molecular size and entanglements on polymer properties.

For example, the Rouse model predicts that the viscosity of a polymer melt depends proportionally to its Mw which is in a relatively good agreement with reported dependency of $M^{1.8}$ for very low Mw polymer melts [32]. However, it has been shown that the viscosity of high Mw polymer melts, where chain entanglement is significant, depends on the Mw by $M^{3.4}$ [33], which is far from the Rouse model prediction.

In order to explain the movement of large polymer molecules, de Gennes [34] proposed the reptation model in which the restriction imposed by neighboring polymers is considered as a temporary tube surrounding the polymer chain. Figure 2.8 schematically shows the tube model surrounding a polymer chain.

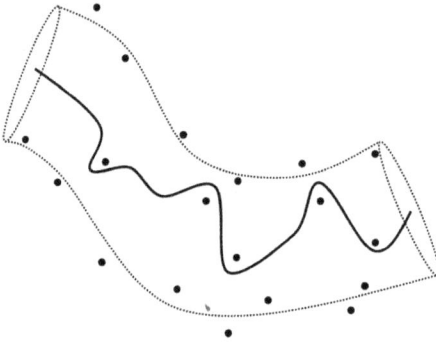

Figure 2.8: Schematic of the de Gennes tube model showing the initial tube around a polymer chain with the surrounding obstacles (shown as black points).

The polymer chain in this model can only diffuse along the axis of the tube. As the chain diffuses, it leaves a part of the initial tube and enters a new tube. This process is called reptation. Based on this model, displacement of polymer molecules over long distances occurs when the polymer chain leaves its initial tube completely. The polymer chain in this model is a Gaussian chain and consists of beads and Hookean springs similar to the one in the Rouse model. Compared to the Rouse model, the reptation model has two additional parameters: the tube diameter and the tube length. Three important relaxation times determine the different regimes of diffusion in the reptation model: (i) the entanglement time (τ_e), (ii) the Rouse relaxation time (τ_R), and (iii) the disentanglement time (τ_d). The entanglement time (τ_e) is the time that a Rouse bead needs to travel the tube diameter to feel constraints of the tube. τ_e is defined as

$$\tau_e = \left(\frac{M_e}{M}\right)^2 \tau_R \tag{2.6}$$

The definition of the Rouse relaxation time (τ_R) is the same as equation (2.3). Finally, the disentanglement time (τ_d) in the reptation model is defined as the required time for a chain to completely leave the initial tube and is defined as

$$\tau_d = 6\left(\frac{M}{M_e}\right)\tau_R \tag{2.7}$$

At times smaller than τ_e, the beads do not feel constraints of the tube; therefore, the mean square displacement of a polymer chain in this region is the same as equation (2.4) and depends on time with $t^{1/2}$. At $\tau_e < t < \tau_R$, the beads move randomly but the whole chain does not move; therefore, the initial tube does not change. The mean square displacement in this regime can be explained by the following equation:

$$<X^2> = \left(\frac{4Db^2}{3\pi}\right)^{1/4} t^{1/4} \tag{2.8}$$

For $\tau_R < t < \tau_d$, the mean displacement is proportional to $t^{1/2}$. Finally, at $t > \tau_d$, the whole chain moves and the mean square displacement of the chain in this region is linearly related to time. Figure 2.9 schematically shows different regimes of molecular interdiffusion described earlier.

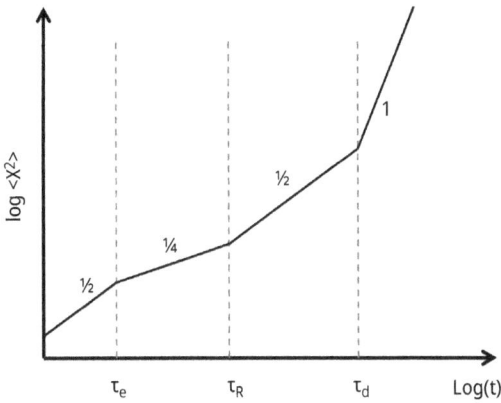

Figure 2.9: Different interdiffusion regimes of a polymer chain in the entangled system based on the reptation model. The shown values are the slope of the lines.

Based on the reptation model, the diffusion coefficient at $t > \tau_d$ is related to Mw by M^{-2}. This relation was later proved by experimental measurements on diffusion of linear polymer chains in the melt state [35–37]. In addition, the reptation model predicts that zero shear viscosity of linear polymer melts with narrow MWD depends on Mw by M^3 which is close to the well-known empirical relation of $\eta_o \sim M^{3.4}$ [32, 33]. These results indicate that the reptation model can successfully predict molecular dynamics of entangled linear polymer chains.

It should be noted that both the Rouse and the reptation models were based on the assumptions of narrow Mw and linear polymer chains. However, many polymers used for sealant layers have broad MWD and short or long chain branches. These parameters have been shown to affect molecular interdiffusion [29, 38, 39] and their effects on heat sealing will be discussed later in Chapter 6.

2.4.2 Interdiffusion at polymer–polymer interfaces

The models described in the previous section were proposed for interdiffusion in the bulk of polymers but in a heat sealing process, diffusion occurs at the interface between two films. Polymer diffusion across the interface has been widely discussed in the literature of welding and healing of polymer–polymer interfaces [23, 40, 41]. The diffusion mechanism at the interface significantly depends on the configuration and situation of the chain ends relative to the interface as well as the free volume at the surface. It is known that the free volume increases by approaching the surface, which increases the diffusion rate at the surface compared to the bulk. The thickness of the surface layer with higher free volume has been estimated to be in the order of 1–10 nm [42]. Wool [25, 43] studied diffusion across the interface of a deuterated PS and a hydrogenated PS surface using secondary ion mass spectroscopy and found that the variation of the interpenetration distance with time follows $t^{1/4}$, which corresponds well with the reptation model prediction. Considering the common dwell times in heat sealing processes and the reptation timescale of high Mw entangled polymer chains, the sealing process is usually done at dwell times much shorter than the disentanglement time (τ_d). The minimum required dwell time for diffusion and entanglement formation in a heat sealing process can be estimated as the time required for polymer chains to diffuse a distance equal to their radius of gyration (R_g). This time is commonly named the interdigitation time and is defined as [44, 45]

$$\tau_i = \frac{R_g^2}{4D_C} \tag{2.9}$$

Qureshi et al. [46, 47] estimated the interdigitation time and diffusion distance for a PE matrix with Mw = 120 kg/mol as 0.04 s and 280 nm, respectively. However, they found that the time required for hot tack strength to reach its maximum value was almost two orders of magnitude greater than the estimated interdigitation time. They attributed this observation to an assumption of the presence of a thin layer of low Mw amorphous chains at the surface of the films. The presence of such amorphous layer on the surface of polymer films has also been reported by some other researchers [48]. Qureshi et al. argued that in order to develop seal strength, large polymer molecules should diffuse through this thin low Mw layer which increases the time required for seal development compared to the estimated interdigitation time.

2.5 Cooling and crystallization

Jaws are opened after dwell time and usually the sealed area begins to cool down by convection heat transfer to the surrounding atmosphere/media. As the temperature of the sealed area decreases, molten polymers in the sealed area begin to solidify and, in the case of polymers with fast crystallization, form dense ordered structures called crystals. The formation of polymer crystals reduces the mobility of polymer chains diffused across the interface and increases the seal strength [1, 3]. The crystallization of semicrystalline polymers from the melt has been the subject of many studies as it directly affects the final properties of polymer matrices, including (but not limited to) mechanical properties [49], moisture [50], and gas permeability [51]. Before reviewing thermodynamics and kinetics of crystallization, first, the structure of a polymer crystal needs to be briefly discussed. It has been shown that in polymer crystals formed from dilute polymer solutions, polymer chain axes are perpendicular to the basal face of crystals. Based on these observations, it was proposed that polymer chains fold in these crystal structures with adjacent reentry into crystals [52]. This model is called the folded chain model and is illustrated in Figure 2.10(a). Flory [53] showed that the adjacent reentry assumption is not valid for crystals formed from polymer melts. He proposed the switchboard model in which polymer chains that leave the basal face of a crystal reenter the crystal but with varying length of the loops, and some chains do not return to the same crystal but enter another crystal and link the crystals. This model is schematically shown in Figure 2.10(b).

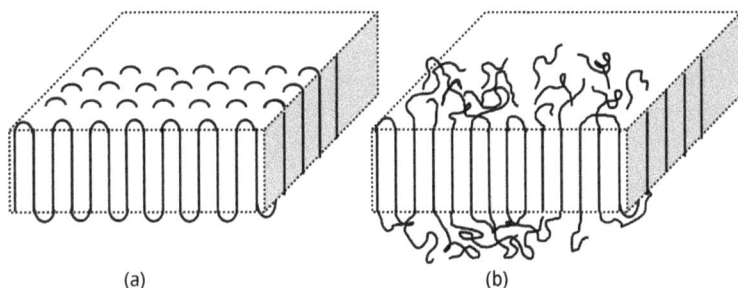

(a)   (b)

Figure 2.10: Polymer crystal models: (a) the folded chain model and (b) the switchboard model.

Crystallization of polymer melts is a nonisothermal process and consists of two main steps: (i) nucleation and (ii) crystal growth. Figure 2.11 shows a typical nonisothermal crystallization rate as a function of temperature for semicrystalline polymers.

As can be seen, by increasing the supercooling degree (which is defined as $\Delta T = T_m - T$), the crystallization rate increases up to a maximum and then decreases gradually. The crystallization process begins with the formation of very small crystals known as primary crystals or crystal nuclei. If the primary crystal is formed without the influence of any foreign particle or surface, the nucleation is called homogeneous nucleation. This type of nucleation is rarely observed in bulk polymers [54]. By contrast, if

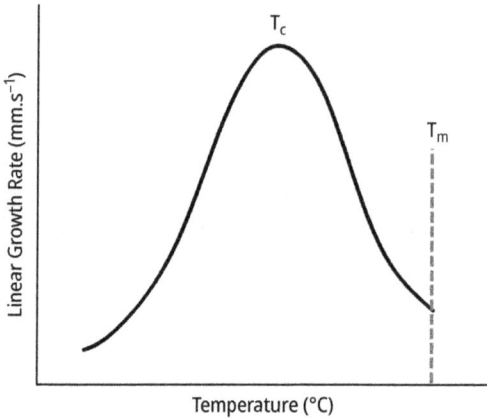

Figure 2.11: Typical crystal growth rate as a function of temperature in nonisothermal crystallization of semicrystalline polymers.

the nuclei are formed on the surface of a foreign particle, the nucleation is called heterogeneous nucleation.

In the case of homogeneous nucleation, the free enthalpy of crystallization (ΔG_C) for a nucleus varies with its size as shown in Figure 2.12. ΔG_C of a homogeneous nucleation is positive for small nuclei and increases by increasing the nuclei size until it reaches a maximum. The nucleus with the maximum ΔG_C is known as a critical nucleus. By further increasing the nucleus size, at a certain size, ΔG_C becomes zero. This certain size is known as nucleus stable size.

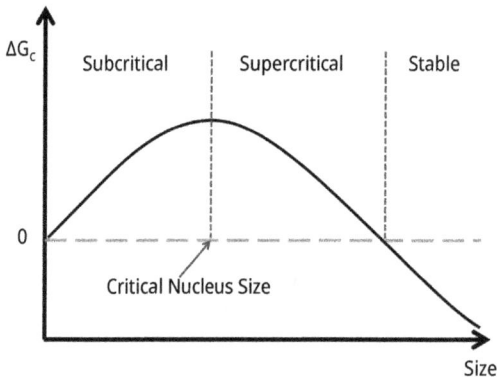

Figure 2.12: Variations of the free enthalpy of crystallization by the nucleus size in homogeneous nucleation.

As shown in Figure 2.12, three different regions of ΔG_C curve indicate three types of nuclei at the beginning stages of crystallization. Nuclei formed with sizes smaller than the critical size are called subcritical nuclei and have the least probability of growth and most likely will be disintegrated. When the nucleus size is between the critical size and the stable size, the formed nuclei have a positive ΔG_C, but increasing the nucleus size reduces the enthalpy of crystallization. These nuclei are much stable and are called supercritical nuclei. Finally, when nuclei size is greater than the stable size, their ΔG_C is negative and they are thermodynamically stable. It has been shown that the enthalpy of crystallization of a primary nucleus with the critical nucleus size, or in other words the energy barrier against nucleation, decreases with supercooling as ΔT^{-2} [55, 56]. As a result, the rate of primary nuclei formation decreases dramatically by approaching the melting temperature. In the case of heterogeneous nucleation, it has been shown that the free enthalpy of crystallization of a heterogeneous nucleus (ΔG_{hetero}) and a homogeneous nucleus (ΔG_{homo}) are related as [57]

$$\Delta G_{hetero} = \Delta G_{homo} \times \phi \tag{2.10}$$

where ϕ can vary between 0 and 1. This relation shows that the energy barrier against the formation of heterogeneous nuclei is lower than that of homogeneous nucleation, which indicates that the heterogeneous nucleation is thermodynamically favored.

After the formation of a stable nucleus, the crystal growth step begins. Hoffman classified the crystal growth into three regimes based on the degree of supercooling [52, 58]. Regime I is at small supercooling and close to the melting temperature. In this regime, one chain is crystallized at the surface of a nucleus at a time. This requires adjacent reentry of the chain. On the other hand, the high temperature of this regime provides high molecular mobility which allows high crystal growth rate. Regime II is observed at moderate supercooling, where multiple polymer chains are crystallized at the same time on the surface of nucleus, with a distance between growth sites known as the niche distance. Hoffman [58] showed that in polyethylene, the transition between regime I and II occurs at a supercooling of 16 °C. Finally, regime III is observed at high supercooling, and the crystallization rate in this regime is very rapid, and the adjacent reentry does not happen, but the formed crystals in this regime show structures similar to the switchboard model. It was shown that in the case of polyethylene, the transition between regime II and III occurs at a supercooling of 23 °C [58]. The crystal growth rate in these regimes can be compared as regime I ~ regime III > regime II. Considering the low nucleation rate in small supercooling region of regime I, regime III is the most desirable regime for crystallization to achieve high crystallinity in industrial applications. As crystallization is an exothermic phenomenon, DSC is widely used to study crystallization of polymers. Figure 2.5 shows schematically the nonisothermal crystallization curve of a polymer material in a DSC test. The crystallization is observed as an exothermic peak in DSC curves. The temperature of the maximum in the crystallization curve is known as the crystallization temperature (T_c). It has been shown by many researchers that increasing the cooling rate shifts the crystallization temperature of polymers to lower temperatures and reduces the enthalpy of crystallization [16, 59–61]. This is attributed to the imperfection induced into the crystal structure at higher cooling rates. In order to analyze the crystallization phenomenon in a heat sealing process, it

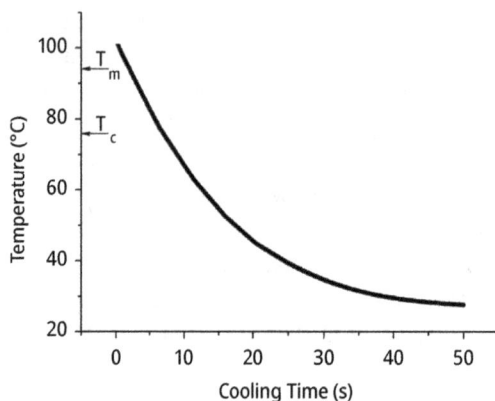

Figure 2.13: Interface temperature as a function of cooling time in a heat sealing process between two PE sealants with $T_{jaw} = 100$ °C. The marked T_m and T_c were determined from DSC analysis at a heating/cooling rate of 10 °C/min (unpublished data).

is necessary to determine the cooling rate of the sealant layer after opening of the jaws. Figure 2.13 shows the estimation of the interface temperature between two polyethylene sealants as a function of cooling time after heat sealing at a jaw temperature of 100 °C.

The interface temperature decreases at an average rate of 3.75 °C/s or 225 °C/min in the first 10 s of the cooling step. Most of the previous works that studied the effects of cooling rate on polymer crystallization were performed using conventional DSC at cooling rates lower than 60 °C/min. This indicates that the effect of high cooling rates used in heat sealing on polymer crystallinity needs to be addressed. It has been shown that the cooling polymer melts at extremely high cooling rates can finally vitrify polymer melts at T_g and form an amorphous polymer [62]. However, this phenomenon happens at cooling rates much higher than the cooling rates in a heat sealing process. For example, it has been shown that cooling HDPE at 60×10^6 °C/min [21, 63] or iPP at 60,000 °C/min [64] prevents crystallization and results in the formation of a completely amorphous polymer. Schawe [12] studied the effect of different cooling rates on crystallization of iPP in a wide range of cooling rates by combining the results of conventional DSC and high cooling rate Flash® DSC. He found that increasing the cooling rates reduced T_c but observed a considerable reduction in the enthalpy of crystallization only at cooling rates greater than 1,800 °C/min. Based on all aforementioned cases, it is expected that experiencing higher cooling rates compared to the one used in conventional DSC analysis of polymers shifts T_c and reduces the enthalpy of crystallization; however, the extent of these effects depends on polymer types.

References

[1]   Stehling, F.C. and P. Meka, *Heat sealing of semicrystalline polymer films. II. Effect of melting distribution on heat-sealing behavior of polyolefins*. Journal of Applied Polymer Science, 1994. **51**(1): p. 105–119.

[2]   Morris, B.A., *Chapter 7: Heat Seal, in the Science and Technology of Flexible Packaging*. 2017, Elsevier: Cambridge.

[3]   Najarzadeh, Z. and A. Ajji, *A novel approach toward the effect of seal process parameters on final seal strength and microstructure of LLDPE*. Journal of Adhesion Science and Technology, 2014. **28**(16): p. 1592–1609.

[4]   Sperling, L.H., *Introduction to Physical Polymer Science*. 4th ed. 2006, John Wiley & Sons, Inc: New Jersey.

[5]   Kanani Aghkand, Z., et al., *Simulation of Heat Transfer in Heat-Sealing of Multilayer Polymeric Films: Effect of Process Parameters and Material Properties*. 2018.

[6]   Tung, L.J. and S. Buckser, *The effects of molecular weight on the crystallinity of polyethylene*. The Journal of Physical Chemistry, 1958. **62**(12): p. 1530–1534.

[7]   Ebewele, R.O., *Thermal Transitions in Polymers, in Polymer Science and Technology*. 2015, CRC: New York.

[8]   Engler, P. and H. Carr Stephen, *Influence of molecular weight on crystallization rate of oriented, glassy nylon 6*. Polymer Engineering & Science, 1979. **19**(11): p. 779–786.

[9]   Brucato, V., et al., *Crystallization of polymer melts under fast cooling. I: Nucleated polyamide 6*. Polymer Engineering & Science, 1991. **31**(19): p. 1411–1416.

[10] Hendra, P.J., et al., *The effect of cooling rate upon the morphology of quenched melts of isotactic polypropylenes*. Polymer, 1984. **25**(6): p. 785–790.

[11] Cavallo, D., et al., *Effect of cooling rate on the crystal/mesophase polymorphism of polyamide 6*. Colloid and Polymer Science, 2011. **289**(9): p. 1073–1079.

[12] Schawe, J.E.K., *Influence of processing conditions on polymer crystallization measured by fast scanning DSC*. Journal of Thermal Analysis and Calorimetry, 2014. **116**(3): p. 1165–1173.

[13] Fereydoon, M., H. Tabatabaei Seyed, and A. Ajji, *Effect of uniaxial stretching on thermal, oxygen barrier, and mechanical properties of polyamide 6 and poly(m-xylene adipamide) nanocomposite films*. Polymer Engineering & Science, 2014. **55**(5): p. 1113–1127.

[14] Fereydoon, M., S.H. Tabatabaei, and A. Ajji, *X-ray and trichroic infrared orientation analyses of uniaxially stretched PA6 and MXD6 nanoclay composite films*. Macromolecules, 2014. **47**(7): p. 2384–2395.

[15] ASTM International, *ASTM D3418-15 Standard Test Method for Transition Temperatures and Enthalpies of Fusion and Crystallization of Polymers by Differential Scanning Calorimetry*. 2015, West Conshohocken: PA.

[16] Liu, T., et al., *Nonisothermal melt and cold crystallization kinetics of poly(aryl ether ether ketone ketone)*. Polymer Engineering & Science, 2004. **37**(3): p. 568–575.

[17] Wurm, A., et al., *Crystallization and homogeneous nucleation kinetics of poly(ε-caprolactone) (PCL) with different molar masses*. Macromolecules, 2012. **45**(9): p. 3816–3828.

[18] Mathot, V., et al., *The Flash DSC 1, a power compensation twin-type, chip-based fast scanning calorimeter (FSC): First findings on polymers*. Thermochimica Acta, 2011. **522**(1): p. 36–45.

[19] Schawe, J.E.K., *Analysis of non-isothermal crystallization during cooling and reorganization during heating of isotactic polypropylene by fast scanning DSC*. Thermochimica Acta, 2015. **603**: p. 85–93.

[20] Hellmuth, E. and B. Wunderlich, *Superheating of linear high-polymer polyethylene crystals*. Journal of Applied Physics, 1965. **36**(10): p. 3039–3044.

[21] Minakov, A.A., A. Wurm, and C. Schick, *Superheating in linear polymers studied by ultrafast nanocalorimetry*. The European Physical Journal E, 2007. **23**(1): p. 43.

[22] Wool, R.P. and K.M. O'Connor, *A theory crack healing in polymers.* Journal of Applied Physics, 1981. **52**(10): p. 5953–5963.

[23] Kim, Y.H. and R.P. Wool, *A theory of healing at a polymer-polymer interface.* Macromolecules, 1983. **16**(7): p. 1115–1120.

[24] Wool, R.P., *Self-healing materials: A review.* Soft Matter, 2008. **4**(3): p. 400.

[25] Wool, R.P., *Polymer Interfaces: Structure and Strength.* 1995: Hanser Publishers.

[26] Brown, H., *The Adhesion Between Polymers*, A.R.M. Sci., Editor. 1991. p. 463–489.

[27] Frederix, C., et al., *Kinetics of the non-isothermal fusion-welding of unlike ethylene copolymers over a wide crystallinity range.* Polymer, 2013. **54**(11): p. 2755–2763.

[28] Zhang, M.Q. and M.Z. Rong, *Theoretical consideration and modeling of self-healing polymers.* Journal of Polymer Science Part B: Polymer Physics, 2012. **50**(4): p. 229–241.

[29] Najarzadeh Z. and A. Ajji, *Role of molecular architecture in interfacial self-adhesion of polyethylene films.* Journal of Plastic Film & Sheeting, 2017. **33**(3): p. 235–261.

[30] Rouse Jr, P.E., *A theory of the linear viscoelastic properties of dilute solutions of coiling polymers.* The Journal of Chemical Physics, 1953. **21**(7): p. 1272–1280.

[31] Bird, R.B., et al., *Dynamics of Polymer Liquids Vol. 2 Kinetic Theory.* 1987, Wiley: New York.

[32] Pearson, D.S., et al., *Viscosity and self-diffusion coefficient of linear polyethylene.* Macromolecules, 1987. **20**(5): p. 1133–1141.

[33] Fox, T.G. and P.J. Flory, *Further studies on the melt viscosity of polyisobutylene.* The Journal of Physical Chemistry, 1951. **55**(2): p. 221–234.

[34] de Gennes, P.G., *Reptation of a polymer chain in the presence of fixed obstacles.* The Journal of Chemical Physics, 1971. **55**(2): p. 572–579.

[35] Bartels, C.R., B. Crist, and W.W. Graessley, *Self-diffusion coefficient in melts of linear polymers: Chain length and temperature dependence for hydrogenated polybutadiene.* Macromolecules, 1984. **17**(12): p. 2702–2708.

[36] Klein, J. and B.J. Briscoe, *The diffusion of long-chain molecules through bulk polyethylene.* Proceedings of the Royal Society of London. A. Mathematical and Physical Sciences, 1979. **365**(1720): p. 53–73.

[37] Antonietti, M., et al., *Diffusion of labeled macromolecules in molten polystyrenes studied by a holographic grating technique.* Macromolecules, 1984. **17**(4): p. 798–802.

[38] Bartels, C.R., et al., *Self-diffusion in branched polymer melts.* Macromolecules, 1986. **19**(3): p. 785–793.

[39] Najarzadeh, Z., A. Ajji, and J.-B. Bruchet, *Interfacial self-adhesion of polyethylene blends: The role of long chain branching and extensional rheology.* Rheologica Acta, 2015. **54**(5): p. 377–389.

[40] Wise, R.J., *Thermal Welding of Polymers.* 1999: Woodhead Publishing.

[41] Prager, S. and M. Tirrell, *The healing process at polymer–polymer interfaces.* The Journal of Chemical Physics, 1981. **75**(10): p. 5194–5198.

[42] Algers, J., et al., *Free volume and density gradients of amorphous polymer surfaces as determined by use of a pulsed low-energy positron lifetime beam and PVT data.* Macromolecules, 2004. **37**(11): p. 4201–4210.

[43] Wool, R., B.L. Yuan, and O. McGarel, *Welding of polymer interfaces.* Polymer Engineering & Science, 1989. **29**(19): p. 1340–1367.

[44] de Gennes, P.G., *Chapter 3 – Mechanical Properties of Polymer Interfaces A2 – Sanchez, Isaac C,* in *Physics of Polymer Surfaces and Interfaces.* 1992, Butterworth-Heinemann: Boston. p. 55–71.

[45] Qureshi, N.Z., et al., *Self-adhesion of polyethylene in the Melt. 2. Comparison of heterogeneous and homogeneous copolymers.* Macromolecules, 2001. **34**(9): p. 3007–3017.

[46] N. Z. Qureshi, et al., *Self-adhesion of polyethylene in the Melt. 1. Heterogeneous copolymers.* Macromolecules 2001(34): p. 1358–1364.

[47] Qureshi, N.Z., et al., *Self-adhesion of polyethylene in the Melt. 1. Heterogeneous copolymers.* Macromolecules, 2001. **34**(5): p. 1358–1364.

[48] Brant, P., et al., *Surface composition of amorphous and crystallizable polyethylene blends as measured by static SIMS*. Macromolecules, 1996. **29**(17): p. 5628–5634.

[49] Landel, R.F. and L.E. Nielsen, *Mechanical Properties of Polymers and Composites*. 1993: CRC press.

[50] Lasoski, S.W. and W.H. Cobbs, *Moisture permeability of polymers I. Role of crystallinity and orientation*. Journal of Polymer Science, 1959. **36**(130): p. 21–33.

[51] Weinkauf, D.H. and D.R. Paul, *Effects of Structural Order on Barrier Properties, in Barrier Polymers and Structures*. 1990, American Chemical Society. p. 60–91.

[52] Sperling, L.H., *The Crystalline State, In Introduction to Physical Polymer Science*. 2006, Wiley.

[53] Flory, P.J., *On the morphology of the crystalline state in polymers*. Journal of the American Chemical Society, 1962. **84**(15): p. 2857–2867.

[54] Binsbergen, F.L., *Natural and artificial heterogeneous nucleation in polymer crystallization*. Journal of Polymer Science: Polymer Symposia, 1977. **59**(1): p. 11–29.

[55] Hu, W., *Intramolecular Crystal Nucleation, in Progress in Understanding of Polymer Crystallization*, G. Reiter and G.R. Strobl, Editors. 2007, Springer.

[56] Lauritzen, J. and J.D. Hoffman, *Theory of formation of polymer crystals with folded chains in dilute solution*. Journal of Research of the National Bureau of Standard A, 1960. **64**(1): p. 73102.

[57] Schmelzer, J.W., *Nucleation Theory and Applications*. 2006, John Wiley & Sons.

[58] Hoffman, J.D., *Regime III crystallization in melt-crystallized polymers: The variable cluster model of chain folding*. Polymer, 1983. **24**(1): p. 3–26.

[59] Ozawa, T., *Kinetics of non-isothermal crystallization*. Polymer, 1971. **12**(3): p. 150–158.

[60] Di Lorenzo, M.L. and C. Silvestre, *Non-isothermal crystallization of polymers*. Progress in Polymer Science, 1999. **24**(6): p. 917–950.

[61] Cebe, P. and S.-D. Hong, *Crystallization behaviour of poly (ether-ether-ketone)*. Polymer, 1986. **27**(8): p. 1183–1192.

[62] Toda, A., R. Androsch, and C. Schick, *Insights into polymer crystallization and melting from fast scanning chip calorimetry*. Polymer, 2016. **91**: p. 239–263.

[63] Minakov, A.A. and C. Schick, *Ultrafast thermal processing and nanocalorimetry at heating and cooling rates up to 1MK/s*. Review of Scientific Instruments, 2007. **78**(7): p. 073902.

[64] De Santis, F., et al., *Scanning nanocalorimetry at high cooling rate of isotactic polypropylene*. Macromolecules, 2006. **39**(7): p. 2562–2567.

Chapter 3
Seal quality and performance evaluation methods

Evaluation of seal area after sealing is of great importance as it provides valuable information about the integrity, quality, and strength of the seal. Presence of micro-size defects could be very difficult and challenging to detect but they can have detrimental effects on packaging performance. For instance, Kirsch [1] showed that the probability of microbial ingress increases significantly when the leak size in nonporous packaging film is greater than only 1 µm. Test methods that are used to examine the seal area in polymer packaging can be generally categorized into two main groups: (i) seal quality and (ii) seal performance tests. The focus of seal quality tests is on detection of any imperfection in the sealed area such as channels and unsealed area. These defects can cause deterioration or contamination of the product. The seal quality tests do not provide any information about the strength of the seal and the results of these tests is binary results of fail or pass. On the other hand, seal performance tests examine the strength of the seal and package integrity. However, in some cases, the results of these tests cannot provide much information about the seal quality and presence of defects in the seal. For example, presence of few micro-channels does not affect seal strength considerably, however, it can cause leaking and contamination of the product. Therefore, seal quality and seal performance tests are complimentary tests and both tests are needed to obtain a comprehensive overview of the seal and package quality/performance.

3.1 Seal quality tests

Before discussing seal quality tests, first, we need to explain different possible defects that may occur in a sealing process. The most possible defects can be listed as follows:

- Unsealed area: unsealed area can be created due to different reasons including (but not limited to) misalignment of package to seal bars, surface contamination of films, presence of external object (product, dust, etc.) between seal sides, partial seal rupture opening due to sterilization, handling, or weight of product.
- Microchannels: fine pathway across the sealed area that can be created due to different reasons including the ones mentioned earlier.
- Narrow seals: This is a phenomenon when the width of the sealed area is reduced at certain part over the whole sealed area. This can be caused by misalignment of the heated bars and the films, due to misalignment of cutter/slitters or due to partial opening of the seal.
- Oversealed area: if the temperature and pressure of sealing machine is too high or the dwell time is long, other layers of the film can also be melted and cause brittle sealing.

https://doi.org/10.1515/9781501524592-003

- Tears/pinholes: it can happen in the material or substrate.
- Microbubbles: bubble formation can happen due to the low and nonuniform seal pressure or due to the presence of contamination at the surface.
- Wrinkle and fold overs: when the film is folded over before sealing, it can form channels in the folded area.

Some examples of these defects are shown below. Figure 3.1(a) shows a microchannel in the seal area which connects two sides of the seal area. Figure 3.1(b) shows a sealed area with unsealed regions in this case due to the presence of a liquid product in the seal area. Figure 3.1(c) shows folding over issue in seal area, and finally Figure 3.1(d) shows a seal with microbubbles in the seal area.

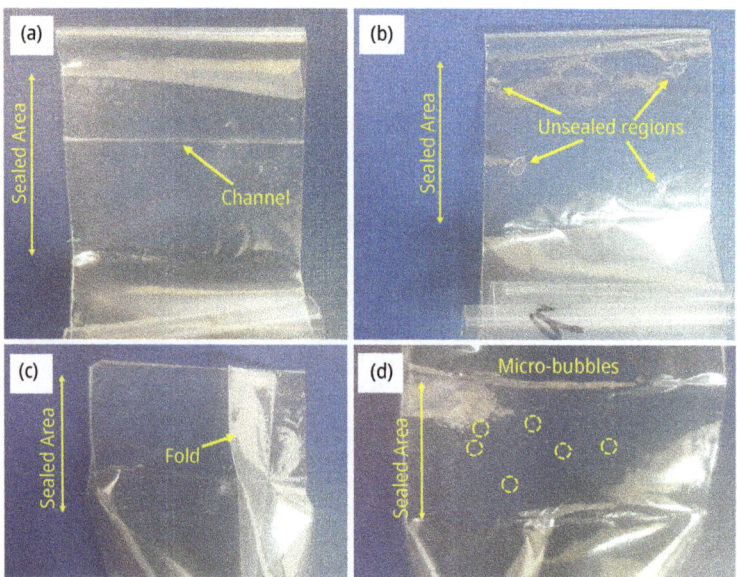

Figure 3.1: Examples of different types of defects in the sealed area.

In the case of channeling, unsealed area and defects that are across the seal area and connect two sides of the seal, leakage rate is a very important parameter that is used to indicate the size of the defect. Different units of leakage rate and their conversion are shown in Table 3.1.

Table 3.1: Seal leakage rate units [2].

Pa.m³/s	Std cm³/s	Std L/day	Air at 0 °C kg/year
1	10	864	400

The most common unit is standard cubic centimeter per second (sccs). From leakage viewpoint, a package is called leak free if the leakage rate is small enough than it will not affect the package quality, protection role and performance. Therefore, mentioning a leak-free package without specifying the limit leakage rate for the desired application is meaningless. Figure 3.2 shows different leakage rates with their practical significance.

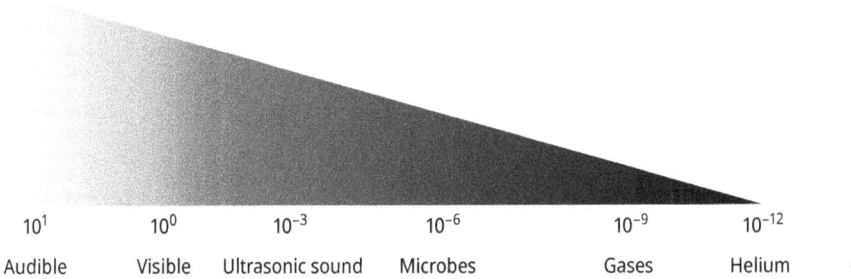

| 10^1 | 10^0 | 10^{-3} | 10^{-6} | 10^{-9} | 10^{-12} |
| Audible | Visible | Ultrasonic sound | Microbes | Gases | Helium |

Figure 3.2: Different leakage rate in sccs units with their significance.

3.1.1 Visual inspection

The first and simplest method for examining seal quality is the visual inspection of the sealed area. This method can be used for channels with width down to 75 μm. ASTM F1886 [3] provides a standard test procedure for visual examining of the seal quality. In this method an illumination of 540 lumen/m^2 is required and the sealed area is examined from 30 to 40 cm distance and the number of defects are recorded.

3.1.2 Flat bar test

This method is a common and handy method for initial examining of the seal area integrity. In this method, a special-shaped flat bar is placed inside the package and is gently forced to the sealed area to make sure no-peeling occurs. The bar is designed so that the curved area such as gusset area and the flat area such as side sealed area can be easily examined. This test provides quick and simple qualitative evaluation of sealed area integrity before doing further tests. The main disadvantage of this testing method is its significant dependency on the pressure applied by the operator which can be different (Figure 3.3).

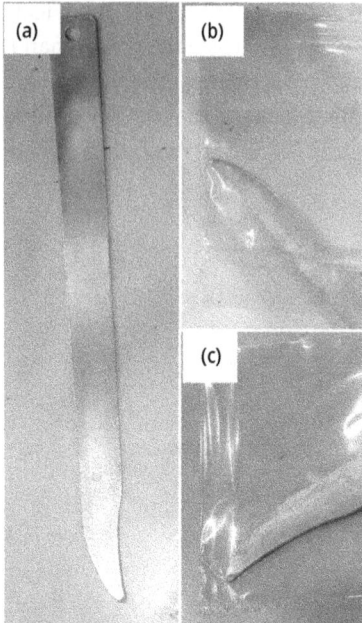

Figure 3.3: Flat curved bar for examination of the sealed area: (a) the bar, (b) examining side seal, and (c) examining corners and gusset seal.

3.1.3 Gross leak or bubble test

This test method is aimed to detect leaks and channels down to 250 µm. According to ASTM F2096 [4], the package is inflated underwater using an internal air pressure introduced into the package by an airway. The package is inflated to a predetermined pressure and then it is checked for air bubbles streams exiting the package. In order to determine the pressure, a defect with known size is created in a package and then the package undergoes inflation. The pressure at which the first air bubble is observed is considered as the minimum inflation pressure that is used in testing. This test method can also be used for vacuum packing products.

In a similar standard test method in ASTM D3078 [5], the package is placed in a sealed water reservoir and then vacuum is applied to the reservoir. The package is inflated under vacuum and air bubble stream can be observed from the leak points/area. This test method can be used for packaging with headspace gas and is not suitable for testing vacuum packaging. Reported laboratory results [5] indicated that high vacuum (~24 in Hg = 600 mm Hg) was required to detect very small leaks (10^{-2} sccs) and those leaks could not be detected in medium vacuum (18 in Hg = 460 mm Hg) that is commonly used to test flexible packages. Similar test technique is used to determine the effect of altitude on packaging [6]. The required vacuum should be determined by creating a leak with known size and then determining the vacuum required to observe air bubbles from that leak. The package is considered to pass the test if no bubble is observed at least for

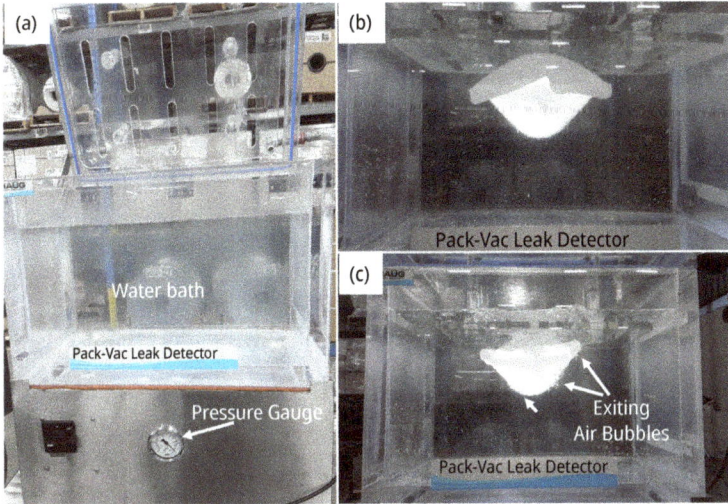

Figure 3.4: Vacuum leak test: (a) testing machine, (b) a package with no leak during the test, and (c) a package with air bubbles emerging from the leaks.

30 s after reaching 18 in Hg. Figure 3.4 shows the testing reservoir and a sample that passed the test (Figure 3.4b) and a sample that failed the test (Figure 3.4c).

3.1.4 Pressure decay leak test

In pressure decay leak test method, the package is pressurized by air to a certain pressure and then air flow is closed and internal pressure of the package is monitored. Decrease in the internal pressure of the package indicates the presence of leaks. This test can be done according to ASTM F2095 [7]. This test method provides a rapid test to detect small leaks in order of 10^{-4} sccs. The following relation can be established to determine the leak rate (Q) from pressure decay results:

$$Q \text{ (sccs)} = \frac{\Delta P \text{ (atm)} \times V \text{ (cm}^3)}{\Delta t \text{ (s)}} \tag{3.1}$$

where ΔP is the pressure decay, V is the package volume, and Δt is test duration. For example, a pressure decay of 1 mbar from a package of 100 cm^3 in 30 s indicates a leak rate of 3.3×10^{-3} sccs.

The test can be done with or without a restraining plate. The restraining plate has two main roles: (i) limits the volume of the inflated bag and therefore reduces test time and increases sensitivity of the test results, (ii) reduces the chance of bursting or peeling the seals. In order to avoid the restraining plate to block the leaks, the restraining plate should be selected from semiporous plastics or surface scored plates or from screen-type materials.

Similar approach is used in ASTM F2338 [8] for vacuum decay leak test in which the package is placed in a chamber and then the chamber is vacuumed. The vacuum valve is then closed, and the chamber pressure is monitored by time. Any increase in chamber pressure indicates leak from the package headspace gas. This test method can detect very small leak rates in order of 10^{-6} sccs and leak sizes down to 5 µm [8].

3.1.5 Dye penetration test

This technique can be used for nonporous packages and can be done in two methods: (A) testing for channels and leaks in the seal or (B) testing for holes or leaks in the flat area. Dye penetration testing procedure is explained in ASTM F3039 [9]. Considering the scope of the book, only the first group of testing is of our interest. This type of testing can be used to detect leaks with the size greater than 50 µm. In this method, a dye penetrant is injected in the package to reach 25 mL of dye per inch (2.54 cm) of the seal area. The injection is usually done from the middle of the package and the package is rotated slowly to make sure the solution is in contact with all sealed edges. The solution remains in contact with the seal area for 5 s. The sealed area in transparent packaging can be examined by eye or by optical device with 5× to 20× magnification. In the opaque packaging, the outside of the sealed area is placed in contact with a dye absorbent.

Due to the contrast between the dye solution and the film, channels, leaks, and unsealed area are shown as lines or strips with the dye solution color. In most sealing companies and converting companies, the dye is placed in the pouch and then the pouch is sealed, and the dye is pushed by hand around the seal area to find out any channel or defect in the sealed area. Figure 3.5 shows an example of a package filled with iodopovidone as the dye and package is examined around the corner and close to the zipper area.

Figure 3.5: Dye penetration test using iodopovidone in a transparent PE zipper bag.

3.1.6 Pressure-assisted dye penetration test

Applying pressure to the package during dye penetration test is a method to reduce the test time and increase the accuracy. In this method, the dye liquid is placed inside the package and then the package is sealed. The sealed package is placed between two plates that apply the pressure. The pressure is applied manually or using a motorized machine to reach the desired pressure. Generally, a pressure around 6 psi or 40 kPa is applied to the package. However, the applied pressure can be increased based on the final application. While the package is kept under the applied pressure, sealed area is examined visually for any dye penetration or leak. Figure 3.6 shows a pressure-assisted dye penetration test in which the pressure is applied by an electronic compression tester.

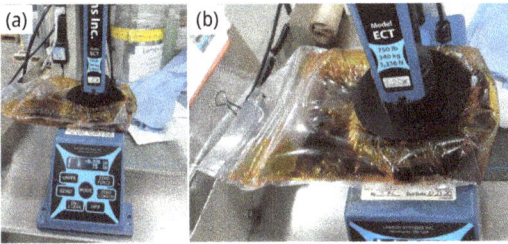

Figure 3.6: Pressure-assisted dye penetration test: (a) the machine and (b) the package under pressure.

3.1.7 Airborne ultrasound

Ultrasound seal quality verification is based on the concept that the ultrasound wave is reflected or absorbed in an unsealed area instead of propagating through the film layers, Figure 3.7. Therefore, defects can be detected by a drop in ultrasound signal received by the detector on the other side of the sealed area.

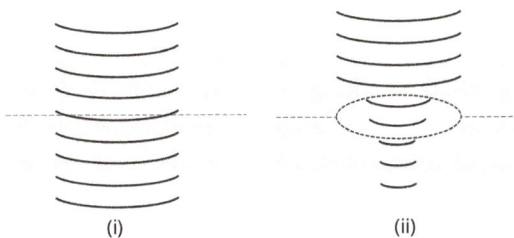

(i)   (ii)

Figure 3.7: Detecting seal defects by airborne ultrasound method: (i) seal without defect and (ii) seal with defect. The curved line represents ultrasound waves. The dashed line shows the interface between two sides of the seal area.

Airborne ultrasound is a noncontact, nondestructive technique that allows analysis of material without contacting to the sensor. The analysis can be done in two modes: linear analysis (L-scan) and 2D analysis (C-scan). The defects are shown typically similar to the one showed in Figure 3.8. ASTM F3004 [10] explains the procedure to perform the seal quality verification using airborne ultrasound. Compared to previous seal quality methods, airborne ultrasound is a much costly technique and requires more training and maintenance costs, but it is the only industrially available technique to monitor seal quality in-line and during sealing process. This unique feature allows higher control on package quality during production and increases considerably the reliability and reduces the production costs especially in the trial step. However, it should be noted that the current airborne ultrasound machines can only be used to monitor seal quality in the sealing lines with line speeds below 30 m/min.

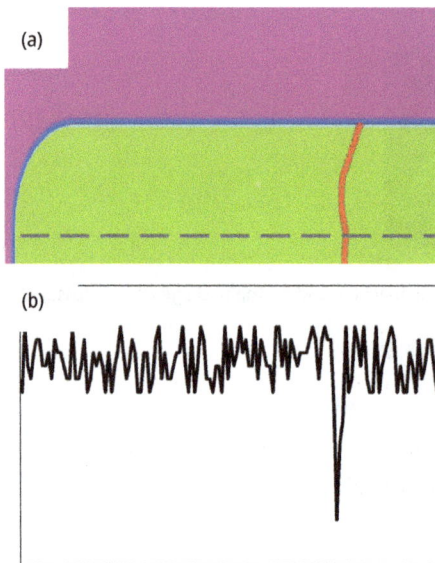

Figure 3.8: Airborne ultrasound scan of a sealed area (a) C-scan and (b) L-scan. The dashed gray line in (a) shows where L-scan was taken. The green area shows the sealed area, pink area is the outside of the package and the red line shows the channel.

3.2 Seal performance tests

Testing techniques used for evaluation of seal properties can be categorized based on the temperature in which the test is done: (i) hot tack tests where the test is done when sealed area is at temperatures above room temperature, and (ii) seal tests where the test is done when the seal area is cooled down to ambient temperature. Both these tests provide the strength of the sealed area as the average required force per unit width of the sealed area to pull apart two sealed films. The first set of test result is called hot tack strength and the second test results are reported as seal strength. The common units of hot tack and seal strength with their conversion coefficients are listed in Table 3.2.

Table 3.2: Conversation factors for seal strength and hot tack units conversion.*

	g/25 mm	g/15 mm	N/m	N/25 mm	N/15 mm	lb/in
g/25 mm	1	1.7	0.4	0.01	0.0163	0.002
g/15 mm	0.6	1	0.235	0.006	0.01	0.0013
N/m	2.55	4.25	1	0.025	0.04	0.0056
N/25 mm	102	170	40	1	1.67	0.225
N/15 mm	61.3	102	24	0.6	1	0.135
lb/in	455	760	180	4.45	7.4	1

*Values are rounded upward to simplify calculations.

Hot tack strength provides important information about seal performance at high temperatures. For example, hot tack is an important factor in vertical form fill seal (VFFS) machines as the package is usually filled when the bottom seal is still hot and should withstand the weight of the product. Hot tack is also important in some horizontal form fill seal (HFFS) applications, for example, when hot tack strength is needed to resist springback forces in gusseted areas (where the films are folded). From industrial viewpoint, achieving required hot tack strength at lower temperatures is desired to increase the production rate and reduce the final cost.

On the other hand, seal strength measured at ambient temperature is an important factor during transportation, storage, and final customer use. Generally hot tack strength is lower than seal strength due to the higher mobility of polymer chains and less crystallinity at higher temperatures which results in entanglement opening and/ or pull out of the chains at lower forces [11]. This point will be explained more in Chapter 6, where the effects of processing conditions on seal performance will be discussed in detail.

It should be noted that based on ASTM, all samples are recommended to be conditioned following ASTM E171 [12] at 23 ± 2 °C and humidity of $50 \pm 5\%$ RH for at least 24 h before testing.

3.2.1 Hot tack test

Hot tack measurement tests are done based on ASTM F1921 [13] standard method. This standard method is restricted for testing by a hot tack machine fulfilling requirements mentioned in ASTM F2029 [14]. The instrument should have a cycle with the

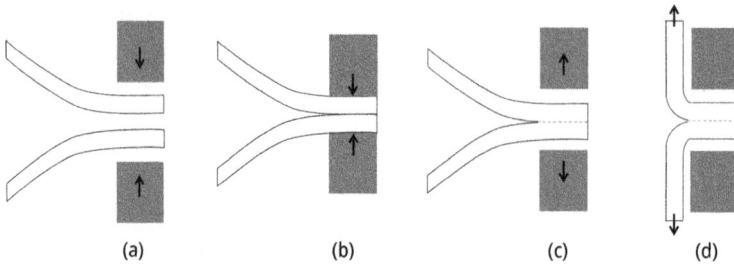

Figure 3.9: Different steps needed in a hot tack test: (a) approaching of the jaws, (b) sealing, (c) delay step, and (d) withdrawal and force measurements. The pointers on the jaws and films show the direction of jaw movement and the applied stress direction to films, respectively.

four following phases: (i) sealing, (ii) delay, (iii) withdrawal, and (iv) force measurement. These testing steps are schematically shown in Figure 3.9.

Based on ASTM F1921, sample strips for hot tack measurements should have length of 25–35 cm (10–14 in) and width of 25 mm (1.00 in) or 15 mm. These sample strips are pressed together between the heated jaws for a predefined dwell time (Figure 3.9(b)). Dwell time should be long enough for the interface between sample strips to reach the desired sealing temperature. A sealing pressure of 40 or 72 psi is common in lab testing but other pressures can also be used. At the end of the sealing phase, the grips move apart after a preset delay time (Figure 3.9(c)). The delay time is defined as the time period between opening of the jaws (end of the sealing phase) and applying the force (the beginning of the withdrawal phase). During the delay time, the seal area is cooled down by heat transfer to the surrounding environment. During withdrawal phase, the machine applies forces to the sample strips to separate them apart ideally at the interface between them. Two hot tack methods were mentioned in the standard method to measure the hot tack strength. The hot tack strength is reported as the maximum force measured during grip separation (withdrawal phase). Figure 3.10 shows a typical shape of the force–distance curve during a hot tack test.

At the beginning of the withdrawal phase, the grips begin to travel apart and stretch the film. The force at this early stage of the withdrawal phase is small until the film is completely straight. At this point, the applied force begins to increase. When the applied force exceeds the hot tack strength, the sealed area begins to be peeled. The peeling continues until the end of the length of the sealed area and drops to zero at the end of the sealed area. According to ASTM F1921 Method B, the maximum applied force recorded is reported as the hot tack strength. In addition to hot tack strength, examining the sample and determining the failure mode is very important in analyzing the results. Figure 3.11 shows the different separation modes during seal test. In the adhesive peeling, the seal is opened under the applied force at the interface between two sealant layers. Adhesive peeling occurs when seal strength is lower than the strength of layers. On the other hand, cohesive peeling is observed

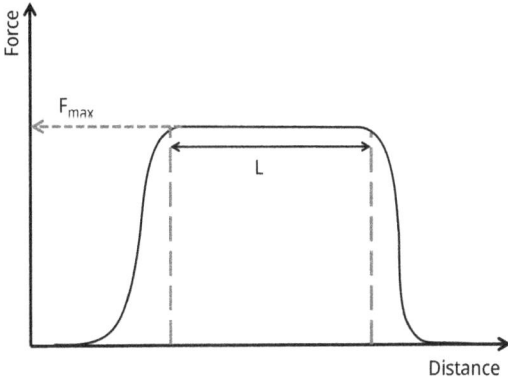

Figure 3.10: A typical hot tack curve showing applied force to the sealed area as a function of the distance traveled by the grips. L is the length of the sealed area and F_{max} is the maximum recorded peeling force.

when hot tack strength is greater than the strength of each sealant layer. This occurs when sealant layer contains some defects that cause stress concentration and crack propagation in the sealant layer. Some common defects could originate from the presence of gel particles, another phase or solid particles present in the sealant layer. The delamination mode is observed when hot tack strength is greater than the adhesion strength between abuse layer (outer layer) and the film. In addition, there are some other modes of failure that can occur during seal test which are shown in Figure 3.12. In the case of break and remote break, the plastic film breaks after applying the force due to the defects in the film or due to the very high strength of the seal. Based on the test results, sealing can be divided into two main groups of peelable and lock seal. Peelable seals are those that show peeling behavior (adhesive or cohesive) during test, while lock seals do not show peeling behavior. Therefore, based on this categorization, seal area showing adhesive peeling, cohesive peeling, and peeling with elongation behavior can be labeled as peelable seals. On the other hand, seals that show delamination, break, remote break, elongation, and tear are categorized as lock seals. Figure 3.13 shows images of the samples with these three types of sealing behavior.

Both samples (a) and (b) show adhesive peeling behavior but sample (a) shows poor hot tack strength and no considerable change in the film appearance after seal opening, while sample (b) shows adhesive peeling behavior with higher seal strength and whitening of the sealed area after peeling. This whitening has been shown to be due to formation of elongated filaments at the seal area during testing [15]. It should be mentioned that distinguishing adhesive and cohesive peel could be challenging in some cases and may require microscopy analysis to determine where peeling occurred. Sample (c) shows very high hot tack strength and tears from sealed area during testing. Based on these results, samples (a) and (b) can be categorized as peelable with adhesive peeling mechanisms and sample (c) as lock seal with tear in sealed area. As hot tack strength depends on the peeling rate [16], it is recommended to examine it at both high (about 12 in/min or

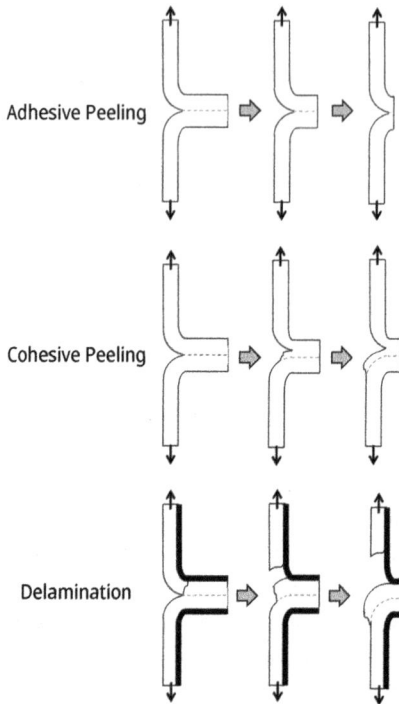

Adhesive Peeling

Cohesive Peeling

Delamination

Figure 3.11: Different modes of seal separation. The dashed lines show the interface between two sealant layers. The dark layers in the last row represent the abuse layer.

5 mm/s) and low (1 in/min or 0.43 mm/s) peeling rates for materials when hot tack strength change considerably with peeling rate. However, in packaging industry, companies use the peeling rate for hot tack test according to their final application. This results in a wide range of peeling rate used in packaging industry, which can result in discrepancies in hot tack strengths of the same sample measured by two different companies.

3.2.2 Seal strength measurement

The typical sample widths for seal strength measurements are 15 and 25 mm. The sealed area length is determined based on the sealer configuration. For example, in heat bar sealing, it is the same as the width of the heated bars/jaws. Seal strength measurements can be done based on ASTM F88 [17] using three different sample supporting options: (i) technique A: unsupported sample, (ii) technique B: supported 90° by hand and (iii) technique C: supported 180°. These three techniques are shown schematically in Figure 3.14.

In technique A, both films are secured in the grips with 90° angle with respect to the unsupported seal area (also referred to as the tail). Due to the unsupported nature of the seal area in this technique, this angle changes during the test. In technique B, the seal area is fixed during the test and remains in 90° angle with respect to the films.

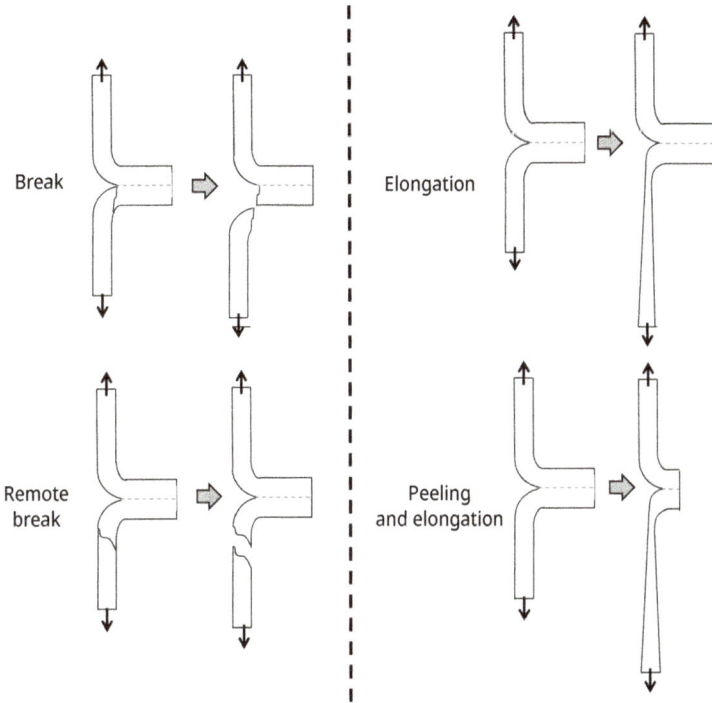

Figure 3.12: Possible failure modes during seal test.

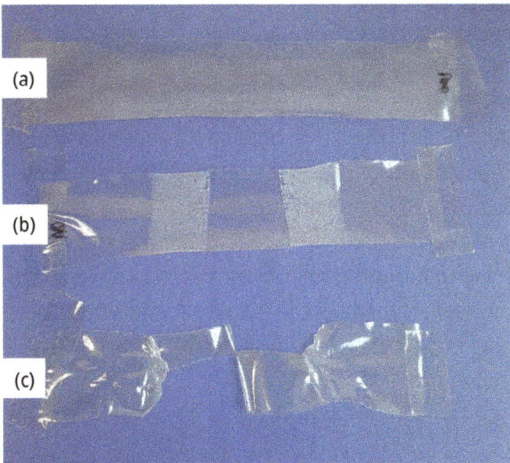

Figure 3.13: Samples showing different modes of separation: (a) adhesive peel (very low seal strength) and (b) adhesive peeling (higher seal strength), (c) tear.

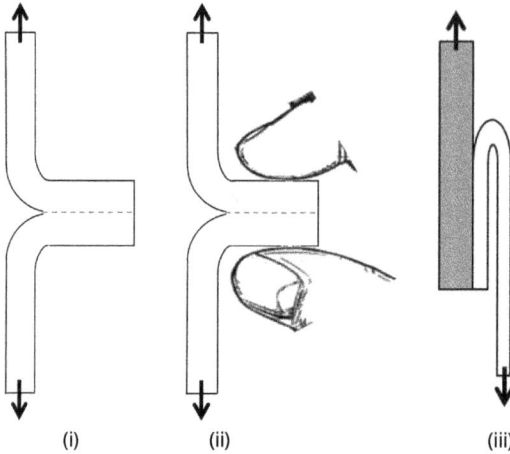

Figure 3.14: Three different techniques used for seal strength test based on ASTM F88: (i) technique A: unsupported, (ii) technique B: supported 90°, and (iii) technique C: unsupported 180°. The pointers show the direction of applied force.

Finally, in technique C, the sealed area is parallel to the films during the test. Technique C is used mostly to evaluate the seal strength of a flexible film to a rigid substrate or sheet. For example, technique C is commonly used to examine the seal strength of lidding films to cups/trays. Methods A and B are commonly used to examine film samples with similar flexibility. It is worth noting that according to ASTM F88, a flexible film is defined as a material for which its flexural strength and thickness allow a turn back to approximately 180°. The steps of the heat seal test are the same as hot tack test shown in Figure 3.9. During the heat seal test, two sides of the sealed area are pulled apart and the applied force opens the sealed area. Seal strength test can be done using a tensile machine or an automated heat seal tester [18, 19]. The result obtained from seal strength test is in the form of the peeling force per width of the sealed area versus the peeled distance (similar to the one shown in Figure 3.10). According to ASTM F88, the reported seal strength should be determined by averaging the force in the plateau region of the curve. However, in practical applications, maximum seal strength is mostly reported, which is defined as the maximum of the seal strength in the plateau region. It is also a common practice in industry to report the maximum strength when the grips traveled a distance equal to the half of the sealed area.

3.2.3 Internal pressurization failure resistance

In this type of testing, the ability of a package to withstand internal pressure is examined. ASTM F1140 [20] discusses three types of testing in this category. In burst test method, the internal pressure is increased continuously until the package fails. The result of this test is the maximum pressure that the package could withstand before failure. Figure 3.15 shows the testing machine to measure burst pressure and Figure 3.16 shows the test result for the shown package. At the early stages of the test, the inside pressure does not change considerably as the introduced air results in inflation of the package. When the package is completely inflated, the pressure increases rapidly until the package fails. The failure should occur in the package walls (such as the one shown in Figure 3.15(c)) and not in the seal area.

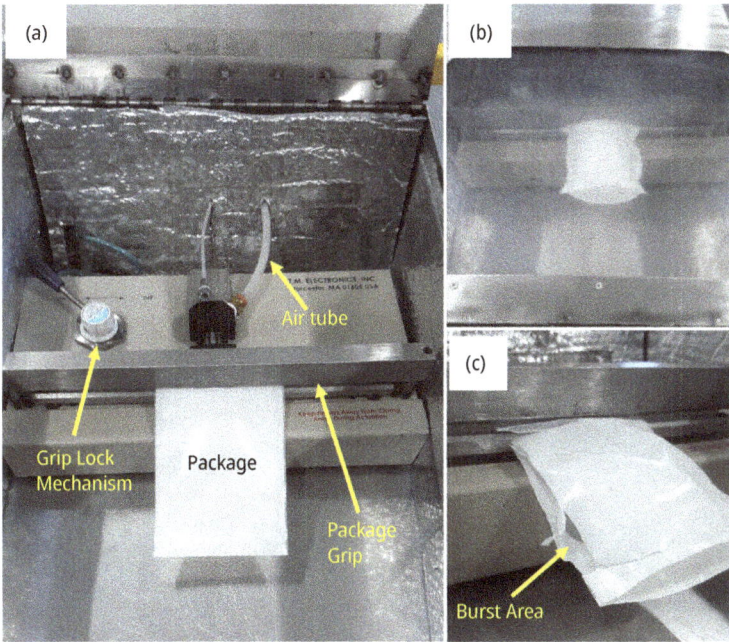

Figure 3.15: (a) Internal pressurization test machine with open package testing, (b) inflated package under the protection cover, and (c) an example of the burst package.

In the creep test method, the pressure is increased to a specific level and then maintained for a certain time. If the package does not fail in the examined time period, it is considered to pass the test. In the third method that is called creep to failure method, the pressure is increased to a specific point and kept constant until the package fails. The result of this test is the time that the package failure is observed.

These types of test can be done for two package types: (i) open package test and (ii) closed packages. The first type is useful to examine seal performance by converting companies that produce pouches for downstream companies that fill the pouch with the product. The second type is mostly used for companies that fill the pouches with products. The instrument that is used for all these tests is the same as the one shown in Figure 3.15(a). However, the configuration for gas inlet is different. In the

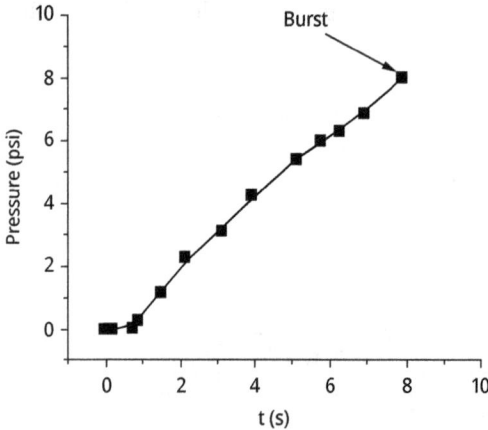

Figure 3.16: Evolution of pressure inside the package during the burst test for the package shown in Figure 3.15.

case of open package testing, gas is introduced from the open side of the package, while in the closed package testing, gas is introduced using a needle inserted in the pouch wall.

References

[1]   Kirsch, L.E., *Package Integrity Testing, in Guide to Microbiological Control in Pharmaceuticals and Medical Devices*, S.P. Denyer and R.M. Baird, Editors. 2006, CRC Press.

[2]   Akers, M.K., D. Larrimore, and D. Guazzo, *Package Integrity Testing, in Parenteral Quality Control: Sterility, Pyrogen, Particulate, and Package Integrity Testing*. 2002, CRC Press.

[3]   Patrício, T., et al., *Fabrication and characterisation of PCL and PCL/PLA scaffolds for tissue engineering*. Rapid Prototyping Journal, 2014.

[4]   Thadavirul, N., P. Pavasant, and P. Supaphol, *Development of polycaprolactone porous scaffolds by combining solvent casting, particulate leaching, and polymer leaching techniques for bone tissue engineering*. Journal of Biomedical Materials Research Part A, 2014. **102**(10): p. 3379–3392.

[5]   *Standard Test Method for Determination of Leaks in Flexible Packaging by Bubble Emission*. 2013.

[6]   Harris, L.D., B.S. Kim, and D.J. Mooney, *Open pore biodegradable matrices formed with gas foaming*. Journal of Biomedical Materials Research: An Official Journal of the Society for Biomaterials, the Japanese Society for Biomaterials, and the Australian Society for Biomaterials, 1998. **42**(3): p. 396–402.

[7]   *Standard Test Methods for Pressure Decay Leak Test for Flexible Packages with and without Restraining Plates*. 2013.

[8]   *Standard Test Method for Nondestructive Detection of Leaks in Packages by Vacuum Decay Method*. 2013.

[9]   Sai, H., et al., *Hierarchical porous polymer scaffolds from block copolymers*. Science, 2013. **341**(6145): p. 530–534.

[10]   Huang, Y., et al., *Preparation and properties of poly (lactide-co-glycolide)(PLGA)/nano-hydroxyapatite (NHA) scaffolds by thermally induced phase separation and rabbit MSCs culture on scaffolds.* Journal of Biomaterials Applications, 2008. **22**(5): p. 409–432.

[11]   Shekhar, A., *A model for hot tack behaiveior in Ethylen Acid co-polymer films.* Tappi, 1994. **77**(1): p. 97–104.

[12]   *Standard Practice for Conditioning and Testing Flexible Barrier Packaging.* 2015.

[13]   *Standard Test Methods for Hot Seal Strength (Hot Tack) of Thermoplastic Polymers and Blends Comprising the Sealing Surfaces of Flexible Webs.* 2018.

[14]   Kurobe, H., et al., *Concise review: Tissue-engineered vascular grafts for cardiac surgery: Past, present, and future.* Stem Cells Translational Medicine, 2012. **1**(7): p. 566–571.

[15]   Najarzadeh, Z. and A. Ajji, *Role of molecular architecture in interfacial self-adhesion of polyethylene films.* Journal of Plastic Film & Sheeting, 2017. **33**(3): p. 235–261.

[16]   Theller, H.W., *Heatsealability of flexible web materials in hot-bar sealing applications.* Plastic Film and Sheeting, 1989. **5**: p. 66–93.

[17]   Giannico, S., et al., *Clinical outcome of 193 extracardiac fontan patients: The first 15 years.* Journal of the American College of Cardiology, 2006. **47**(10): p. 2065–2073.

[18]   Fiesser, F.H. and R.V. Jeral, Transverse heat-sealing apparatus for continuous shrink film packaging, US5475964 A, 1995.

[19]   Montano, L.M., J.M. Montano, and W.J. Witzler, Method and apparatus for testing the quality of crimped seals, US7179344 B1, 2007.

[20]   Kim, T.K., et al., *Gas foamed open porous biodegradable polymeric microspheres.* Biomaterials, 2006. **27**(2): p. 152–159.

Chapter 4
Sealant layer materials

This chapter is dedicated to a summary of the most common sealant materials that are used in plastic packaging. It is worth mentioning that only pure materials are reviewed here and multiphase sealant materials are presented in Chapter 7 as some theoretical concepts need to be introduced first.

4.1 Polyethylene

Polyethylene is the most common sealant layer material used in plastic packaging industry due to its low cost, availability, ease of processing, good seal performance, high moisture barrier, high thermal stability, and chemical resistance. The repeat unit for polyethylene is shown in Figure 4.1.

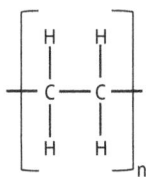

Figure 4.1: Repeat unit of polyethylene.

As can be seen, polyethylene has a saturated structure without any functional group which can explain its high chemical and thermal stability [1]. Polyethylene can only be dissolved in non-polar solvents such as cyclohexane at high temperatures [2]. This indicates the significant potential of polyethylene as the inner layer in contact with chemicals or acidic foods. It is worth mentioning that the chemical resistance and barrier properties should not be confused here. Chemical resistance indicates how much a solvent can change the nature of the material by dissolution or degradation, while barrier properties indicate the resistance of the material against permeation of a substance. For example, polyethylene has good resistance against chemicals and is one of the best choices for the sealant for detergent applications, but its barrier properties are not enough to ensure inhibiting chemicals permeation out of the package. Therefore, a high barrier layer or coating is usually used within these multilayer structures to prevent migration of the substances in these products. In addition, the lack of functional groups in polyethylene structure results in a nonpolar hydrophobic surface [2], which is suitable for applications where low adhesion to polar liquids is desired. Figure 4.2 shows schematically the most common categorization of different polyethylene types based on their molecular architecture.

https://doi.org/10.1515/9781501524592-004

(i)

(ii)

(iii)

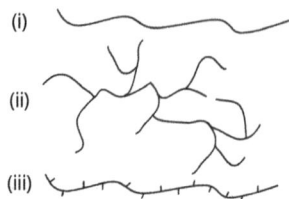

Figure 4.2: Different molecular architectures of polyethylene: (i) linear chain with no or very few side groups, (ii) long-chain branched, and (iii) linear chain with side groups.

4.1.1 High-density polyethylene (HDPE)

Polyethylene with linear chain structure (Figure 4.2(a)) is known as the high-density polyethylene or HDPE. The linear molecular structure of HDPE allows high level of molecular arrangement and leads to a highly crystalline material. Due to the high crystallinity of HDPE, it commonly has the highest density (0.945–0.97 g/cm), melting temperature (130–140 °C), water vapor barrier, oxygen barrier, and mechanical strength among different types of polyethylene. On the other hand, the high crystallinity of HDPE dramatically reduces its clarity. Previous studies showed that the low clarity of HDPE film is mainly originated from light reflection at the film surface due to film surface asperities [3]. HDPE is commonly used in plastic packaging to provide mechanical strength and thermal resistance. Due to its high melting temperature, HDPE is not a common material for the sealant layer in packaging applications. However, HDPE liners are commonly used as the interior layer of multiwall paper bags to provide moisture barrier and allow sealing of the bags. In those applications, ultrasonic sealing is a very common sealing method that allows quick sealing without burning of the exterior kraft paper layers. Figure 4.3 shows the typical footprint peaks of a polyethylene film in Fourier transform infrared (FTIR). A small peak observed at 1,735–1,750 cm^{-1} is assigned to stretching of the C=O band of antioxidants that are commonly added to commercial polyethylene during pelletizing process in petrochemical plants.

The peak at 720 cm^{-1} is assigned to CH$_2$ rocking and is observed for CH$_2$ sequences with a length greater than 4. The peaks at 1,465, 2,850, and 2,920 cm^{-1} are related to CH$_2$ bending, CH$_2$ symmetric stretch, and CH$_2$ asymmetric stretch, respectively [4].

4.1.2 Low-density polyethylene (LDPE)

Low-density polyethylene (LDPE) is produced by free radical polymerization of ethylene in high-pressure tubular reactors [5]. This process results in a polymer with very long chain branches (LCB, Figure 4.2(ii)) that can have sizes comparable to the backbone length. It should be noted that in the classification of side-chain branches, branches with more than 100 repeat units are referred to as LCB which can form entanglement themselves [6]. The presence of LCB results in poor molecular order and reduces the melting temperature, crystallinity, and density of LDPE compared to HDPE. Consequently, LDPE typically shows

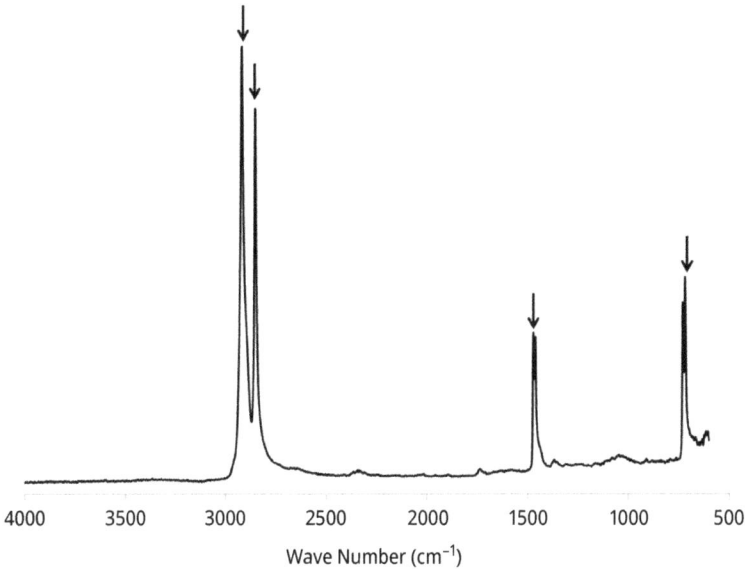

Figure 4.3: Typical FTIR spectrum of polyethylene.

a melting temperature in the range of 100–115 °C and a density of 0.91–0.925 g/cm³. In addition, LCB in LDPE structure suppress its molecular interdiffusion and results in high sealing temperature and poor seal performance [6]. On the other hand, the presence of LCB enhances significantly the melt strength and stability of LDPE in film processing, especially in blown film applications [7]. Therefore, optimizing LCB density in order to achieve a balance between processing and sealing is essential in LDPE.

4.1.3 Linear low-density polyethylene (LLDPE)

Figure 4.4 shows the molecular structure of linear low-density polyethylene (LLDPE) repeat unit. The side-chain groups in LLDPE, which are referred to as short-chain branches (SCB) in some literature, are produced by random copolymerization of ethylene and

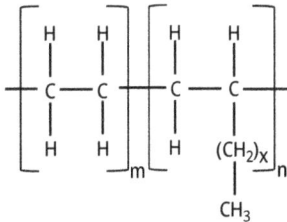

Figure 4.4: Molecular structure of LLDPE with ethylene (left) and alpha olefin (right) sequences.

another alpha olefin comonomer. Comonomers of 1-butene, 1-hexane, and 1-octene are the most common alpha olefin comonomers used in polymerization of LLDPE.

The presence of side groups in LLDPE repeat unit results in the appearance of a small methyl deformation band peak in the range of 1,378–1,383 cm^{-1} in FTIR spectrum of LLDPE. This is very close to the small CH_2 wagging peak located at 1,368 cm^{-1} in the amorphous phase of polyethylene homopolymer. The side groups in LLDPE prevent high level of chain order and reduce its melting temperature and crystallinity compared with HDPE. Consequently, density (0.91–0.93 g/cm) and melting temperature of LLDPE (120–125 °C) are lower than HDPE. Figure 4.5 shows a typical dynamic scanning calorimetry (DSC) heating curve of an LLDPE resin.

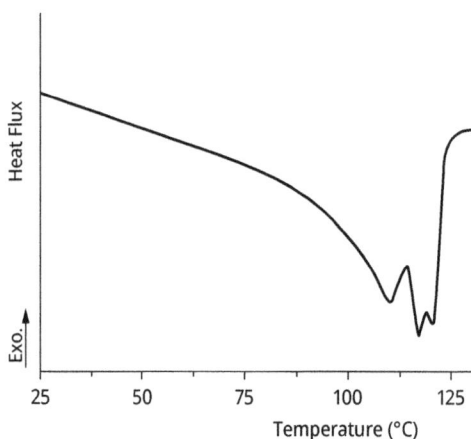

Figure 4.5: DSC heating curve of an LLDPE showing peaks at 109, 119, and 122 °C.

The DSC results show three distinct melting peaks 109, 119, and 122 °C. Although this result may imply that LLDPE has three types of crystals that melt at different temperatures, but previous works used temperature-modulated DSC and showed that the three peaks are observed due to the overlap of multiple exothermic crystal perfection phenomena with endothermic crystal melting [5].

The lower crystallinity of LLDPE results in high clarity and toughness and makes LLDPE a great candidate for many film applications. The presence of side groups in LLDPE delays molecular diffusion but due to their small sizes, the effect is much less pronounced compared to that of LCB in LDPE [8]. The effect of side chains and molecular architecture on seal performance is discussed later in detail in Chapter 5. However, the lack of LCB and low melt strength lead to instabilites in LLDPE blown film production and it is a common practice to blend 10–20 wt% of LDPE with LLDPE to enhance their melt strength and processing. The effect of addition of LDPE on thermal properties of LLDPE was examined by Prasad [9], who showed that the addition of

LDPE to Bu-LLDPE and Oct-LLDPE increases crystal perfection at high temperature but did not have a significant effect on Hex-LLDPE. At the same time, addition of LDPE can delay seal initiation when LLDPE is the sealant material.

4.1.4 Metallocene polyethylene (mPE)

Metallocene polyethylene (mPE) resins are a family of polyethylene that are produced using metallocene catalyst family and can offer interesting properties due to the much higher control on the chain microstructure compared to the conventional Ziegler–Natta polymerization. This leads to much narrower molecular weight distribution (MWD) in mPE and allows tailoring properties to achieve high toughness and puncture resistance and superior heat sealing performance with high hot tack strength. Figure 4.6 shows schematically the difference between LLDPE produced by metallocene catalysts and conventional Ziegler–Natta catalysts. As can be seen, chains produced by metallocene catalysts have almost identical lengths with an even distribution of side groups within each chain and similar side group density among different chains. However, chains produced by Ziegler–Natta catalysts have a broad MWD with an uneven side group distribution within each chain and different side group density among different chains.

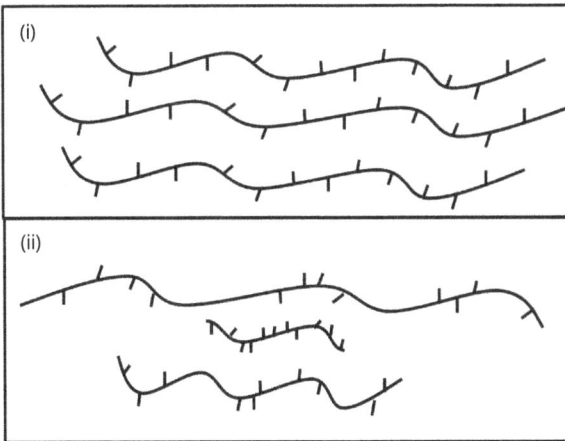

Figure 4.6: Different molecular structures of LLDPE chains produced by (i) metallocene catalyst and (ii) Ziegler–Natta catalyst.

Different types of polyethylene can be produced by metallocene polymerization but the most interesting category of mPE in flexible packaging is that of copolymers of ethylene and alpha olefins or m-LLDPE. Among alpha olefins, 1-hexane and 1-octane are the most common types of mPE that exist in the market. mPE with different densities from <0.9 to 0.935 g/cm^3 can be found in the market where their melting temperature increases with density. Despite their interesting properties, mPE resins are more expensive compared to conventional Ziegler–Natta LLDPE resins. In order to achieve a balance between properties and costs, mPE resins are used as a thin sealant layer or are blended with regular LLDPE resins. The effect of blending m-LLDPE with conventional LLDPE is discussed in detail in Chapter 7.

4.1.5 Plastomers

Plastomer is a group of ethylene and alpha olefins copolymers having a unique molecular structure that allows them to have the performance of rubber-like materials while they can be processed as thermoplastics. The recent advances in catalyst technology allowed producing LLDPE resins with a combination of LCB and SCB in their microstructure. The presence of LCB enhances melt strength and blown film processing while SCB leads to lower crystallinity, higher toughness, and more sticky material and ensures enhanced molecular mobility and interdiffusion. The ratio of LCB/SCB in plastomer resins dictates their final properties and processing. For example, resins with low level of LCB have much higher toughness and good sealing properties but are more difficult for processing. On the other hand, resins with high level of LCB have good processing but lack good sealing and toughness. A detailed discussion on the effect of molecular architecture on seal performance is presented in Chapter 5.

4.2 Polypropylene (PP)

Polypropylene (PP) is produced by polymerization of propylene using Ziegler–Natta or metallocene catalysts in gas phase or slurry reactors [10]. The molecular structure of PP is shown in Figure 4.7. Similar to polyethylene, PP does not have any functional group and has a hydrophobic nature. However, the presence of a tertiary carbon in PP structure reduces its resistance against photo and thermal oxidation compared to PE [11].

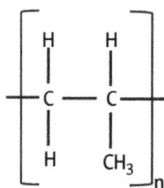

Figure 4.7: Molecular structure of polypropylene.

FTIR spectrum of PP (Figure 4.8) is known for its finger-like peaks in 2,800–3,100 cm^{-1} range.

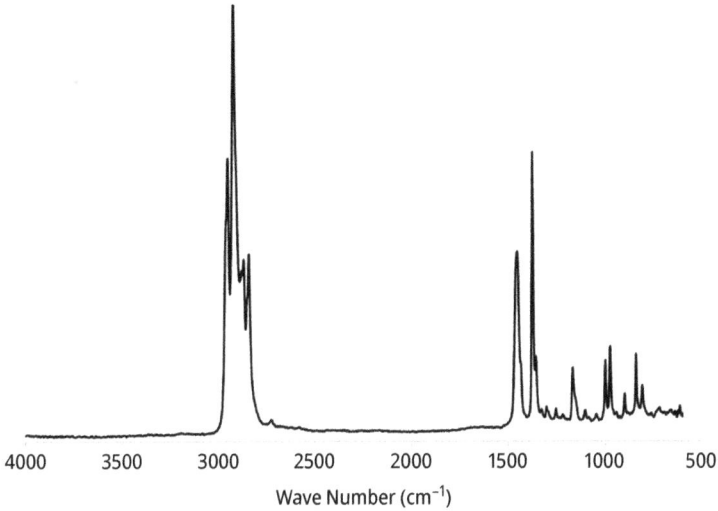

Figure 4.8: FTIR spectrum of polypropylene.

PP can be categorized into three different types based on their tacticity: (i) isotactic (i-PP), (ii) syndiotactic (s-PP), and (iii) atactic PP (a-PP). Figure 4.9 shows schematically the difference between tacticity of these structures.

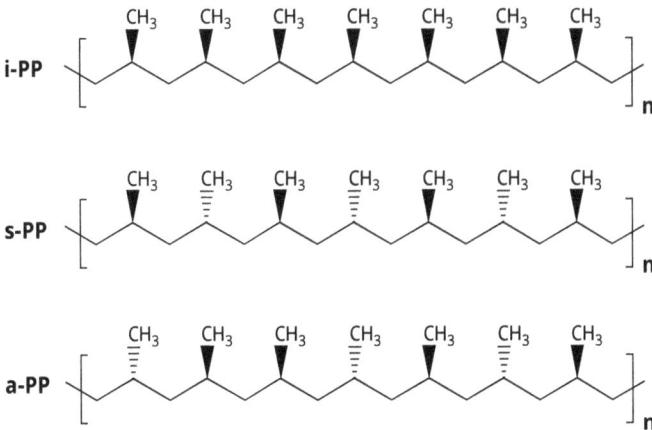

Figure 4.9: Different types of polypropylene based on their tacticity.

As in i-PP all methyl groups are located on one side of the polymer chain, a high level of molecular order and crystallinity can be achieved, which leads to the highest crystallinity and melting temperature compared to s-PP and a-PP. Due to the higher crystallinity and mechanical strength, i-PP is the most commonly used type of PP, particularly in rigid plastic packaging. Industrial grades of i-PP have usually melting temperatures around 155–160 °C. The high melting temperature of i-PP makes it a great candidate for applications where the film needs to withstand high temperatures such as retort applications or microwavable packages. However, high crystallinity of i-PP leads to low clarity of the film. Clarifiers are commonly added to i-PP to improve its clarity. These additives act as nucleating agent and reduce the size of the spherulites in i-PP leading to much clearer films. In s-PP, methyl groups are alternatively located on each side of the chain, resulting in a higher steric hindrance compared to i-PP and therefore lower crystallinity. s-PP has usually melting point around 130 °C and exhibits poor mechanical strength. In a-PP structure, methyl groups are randomly located on either side of the chain which imposes a significant steric hindrance against molecular arrangement. Consequently, a-PP is an amorphous polymer soluble in a great number of aliphatic and aromatic hydrocarbons, esters, and other solvents [12]. The level of tacticity in PP is represented by isotactic index which is determined as the insoluble fraction of PP in boiling heptane. The commercial grades of PP have an isotactic index between 85% and 95%.

PP films that are commonly used in packaging industry are classified based on their orientation level as cast PP (cPP), monoaxially oriented PP (MOPP), and biaxially oriented PP (BOPP) with their level of molecular orientation following the order: cPP < MOPP < BOPP. In fact, cPP represents PP films that did not undergo any postprocessing stretching while MOPP and BOPP went through postprocessing stretching in uniaxial and biaxial directions, respectively. These processes increase considerably the level of molecular orientation in the films. Previous studies showed that increasing molecular orientation reduces the seal strength in PP films [13]. In order to achieve both good strength and sealability, BOPP films usually have three-layer structure of A/B/A, where the A layer is the copolymer of PP and polyethylene and the B layer is PP homopolymer. Due to their low orientation potential and melting temperature, the skin layers provide sealability, while the core homopolymer layer provides strength.

4.3 Ethylene vinyl acetate (EVA)

Ethylene vinyl acetate (EVA) is a random copolymer of ethylene and vinyl acetate (VA) produced by radical polymerization in tubular or continuous stirred reactors. Figure 4.10 shows the molecular structure of EVA.

The presence of VA groups in EVA disrupts crystallinity of ethylene chains and results in a less crystalline material. This increases tackiness, toughness, and clarity while reduces the mechanical strength [14]. In addition, the presence of VA in the structure allows better adhesion and interaction with polar components and surfaces. VA monomer can act as transfer agent in the polymerization reaction which leads to reduced MW of EVA at higher VA contents. EVA is characterized by the amount of VA used in the copolymer. The most common VA contents in plastic packaging are 9% and 18% but higher VA contents also exist and used in some other applications such as hot

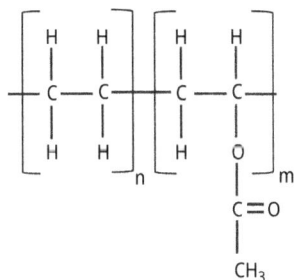

Figure 4.10: Molecular structure of EVA.

melt adhesives. The higher VA content in EVA results in larger fingerprint peaks at wave numbers of 1,020 (C–O band), 1,235 (C–O band), and 1,735 (C=O band). Figure 4.11 shows the FTIR spectrum of an EVA with 9% VA content. It is worth mentioning that the peak around 3,400 cm^{-1} is due to humidity absorption.

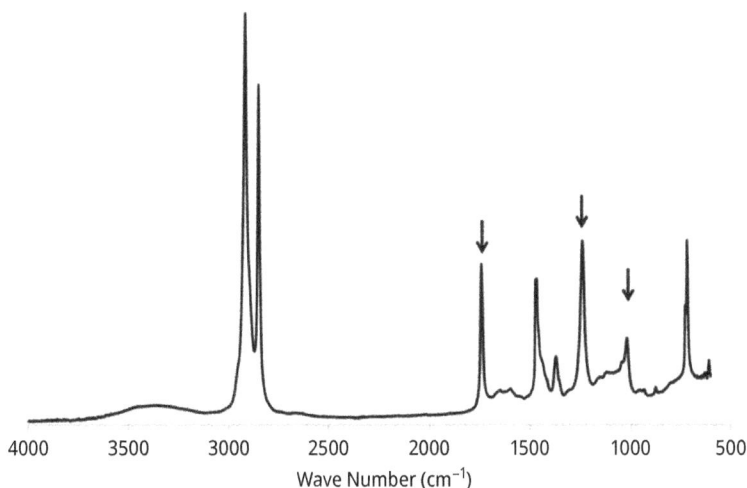

Figure 4.11: FTIR spectrum of EVA with 9% VA content.

In contrast with polyethylene, the density of EVA is controlled by its VA content and increases linearly with an increase in the VA content. For example, Henderson [14] reported the following equation for the variation of density of EVA with VA content:

$$\text{Density} \left(\text{g/cm}^3 \right) = 125 \times 10^{-5} \times \text{VA content} \left(\% \right) + 0.915$$

For example, this equation leads to a density of 0.937 g/cm^3 for EVA with 18% VA content. Increasing VA content also reduces dramatically the stiffness of EVA and results in transition from thermoplastic behavior at 5% VA content to elastomeric behavior at 17% VA content [14].

EVA has been used in plastic packaging for two main reasons: its elastic properties and low sealing temperature. EVA used to be the main option for sealing of heat sensitive materials like cheese and chocolates. Blending of EVA and conventional PE has been considered as an effective approach to reduce sealing temperature of conventional PE sealants. An example of the effect of blending EVA with PE on seal properties is presented in Chapter 7. The invention of mPE, plastomers, acid copolymers, and ionomers reduced considerably the applications of EVA as a sealant material due to their better seal properties, less challenging processing, and their competitive cost (especially mPE and plastomers).

4.4 Acid copolymers

Copolymers of ethylene and methyl acrylic acid (EMA) or ethylene and acrylic acid (EAA) are the most well-known acid copolymers used in packaging industry. The molecular structure of these copolymers is shown in Figure 4.12.

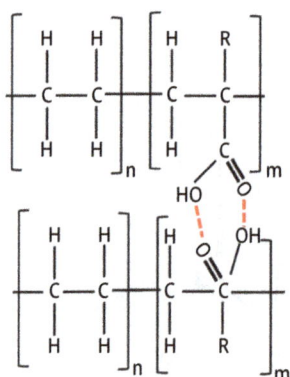

Figure 4.12: Molecular structure of acid copolymers, R is hydrogen in ethylene-acrylic acid (EAA) copolymer, while R is CH_3 in ethylene–methacrylic acid (EMA) copolymer. The red lines show the possible hydrogen bonding between acid moieties.

Copolymerization of ethylene with acid copolymers allows combining water resistance and chemical resistance of ethylene with adhesion and polarity of acid segments. The presence of acid comonomers increases density and improves low temperature sealing and hot tack strength. Figure 4.13 shows the FTIR spectra of EMA (10% MA) and EAA (6.9%AA).

The peak at 1,705 cm^{-1} is assigned to the C=O stretch in dimerized acid and the peak at 940 cm^{-1} is related to the O–H bending in the dimer structure. The peak at 940 cm^{-1} is used commonly to examine relative acid comonomer content in EMA and EAA copolymers. The presence of acid comonomers allows the formation of hydrogen bonding between acid moieties (Figure 4.12), which leads to the appearance of a shoulder-like peak at 2,650 cm^{-1}. This strong intermolecular interaction allows high melt

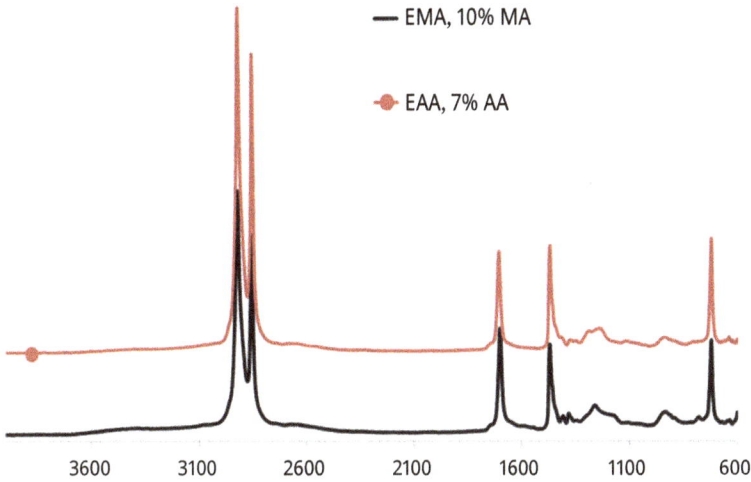

Figure 4.13: FTIR spectra of acid copolymers: EMA with 10% MA and EAA with 7% AA.

strength and high hot tack strength. It has been shown that increasing acid comonomer increases modulus by increasing the glass transition temperature of the amorphous phase [15].

Acid copolymers, especially EAA copolymers, are known for their great adhesion to aluminum foil which makes them a great bonding layer between polyolefin layers and aluminum foil in extrusion coating processes. But processing of acid copolymers has its own challenges as it requires corrosion-protected extrusion equipments. Acid copolymers are tacky materials which makes working with them challenging especially in blown film extrusion processes. In addition, they need to be purged after extrusion at shutdown as they can be cross-linked if exposed to heat for long period of time.

4.5 Ionomers

Ionomers are ion-containing polymers that exhibit interesting properties due to the presence of the ionic group in their structure. Ionomers can be categorized based on the nature of the ion group: (i) anionomers (ionomers with anionic group), (ii) cationomers (ionomers with cationic group), and (iii) polyampholytes and zwitterionomers (ionomers with both anionic and cationic moieties). The majority of ionomers in packaging are anionomers and more specifically anionomers with carboxylate groups that are partially or completely neutralized with alkali metals or zinc. Figure 4.14 shows schematically the structure of zinc ionomers based on polyethylene.

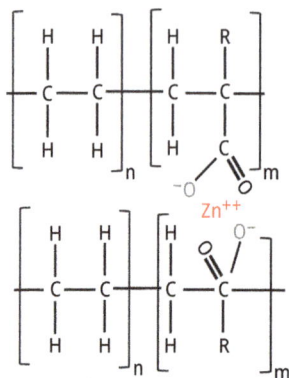

Figure 4.14: Molecular structure of zinc ionomer.

As ionomers are composed of a nonpolar (e.g., ethylene or styrene) segment and a polar (ion) segment, their morphology has been the subject of many studies, and different models have been proposed for their microstructure. It has been shown that incorporation of ionic group changes the molecular morphology in different scales from nano- to microscales [16]. The transition in structures of ionomers below and above melting temperature of their semicrystalline matrix (e.g., ethylene segments) is shown schematically in Figure 4.15.

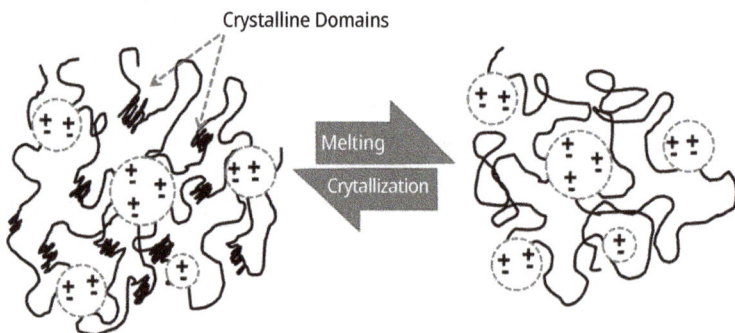

Figure 4.15: Structural transition in ionomers at melting temperature of their semicrystalline matrix.

By increasing temperature, crystalline domains of the polyethylene phase melt but the ionic domains remain intact. These ionic domains act as cross-links points and hinder chain movement which results in increasing melt strength in ionomers. It is worth mentioning that some previous studies considered another order–disorder transition within the ionic domains at temperatures below melting of the semicrystalline matrix [17] which has been debated by some other researchers [18]. Previous studies showed that the rheological behavior of ionomers does not obey the Cox–Merz law. This confirms the presence of two-phase structural model where ionic domains are embedded in a matrix of hydrocarbons [19]. In FTIR spectra of anionomers with carboxylate group, the peak around 1,560 cm^{-1} is assigned to carboxylate ions and its intensity increases by increasing the extent of neutralization [20].

In plastic packaging, ionomers are known for their fat and grease resistance, their low seal temperature, their superior hot tack, and their seal through properties [21]. All these features have made ionomers an interesting candidate where high-quality hermetic sealing is necessary such as in pharmaceutical applications. However, ionomers are known to be relatively expensive, moisture sensitive, and their processing is challenging due to their tackiness and temperature sensitivity.

4.6 Polyethylene terephthalate (PET)

The molecular structure of polyethylene terephthalate (PET) is shown in Figure 4.16. The presence of the aromatic ring in the backbone of PET increases considerably the chain stiffness and strength of PET. In addition, the stiff backbone of PET allows achieving high level of molecular orientation upon stretching. The presence of polar groups in the microstructure of PET provides good interfacial adhesion with inks and good printability. All these properties make PET a promising option for the external abuse layer that provides mechanical strength and printability. The most common form of PET used in the abuse layer in flexible packaging is biaxially oriented PET or BOPET.

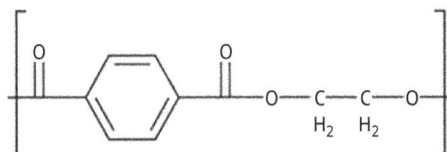

Figure 4.16: Molecular structure of PET.

PET is industrially produced by condensation polymerization of ethylene glycol and terephthalic acid in the presence of a catalyst [22]. Due to its high melting temperature, crystalline PET is not used as a sealant layer; however, amorphous PET (A-PET) is commonly used for lidding of trays made of either cast PET or A-PET. Transparent PET-based film for lidding applications are two-layer films with a thicker BOPET layer as an abuse layer and a thinner A-PET layer on the sealing side. The seal initiation temperature for PET lidding films is commonly 90–100 °C.

References

[1]   Woishnis, W. and S. Ebnesajjad, *Chemical Resistance of Thermoplastics*. 2011, William Andrew.
[2]   Jalali Dil, E. and B.D. Favis, *Localization of micro and nano- silica particles in a high interfacial tension poly(lactic acid)/low density polyethylene system*. Polymer, 2015. **77**: p. 156–166.

[3]   Ashizawa, H., J.E. Spruiell, and J.L. White, *An investigation of optical clarity and crystalline orientation in polyethylene tubular film*. Polymer Engineering & Science, 1984. **24**(13): p. 1035–1042.

[4]   Noda, I., et al., *Group Frequency Assignments for Major Infrared Bands Observed in Common Synthetic Polymers, in Physical Properties of Polymers Handbook*, J.E. Mark, Editor. 2007, Springer: New York, NY. p. 395–406.

[5]   Peacock, A., *Handbook of Polyethylene: Structures: Properties, and Applications*. 2000, CRC press.

[6]   Najarzadeh, Z. and A. Ajji, *Role of molecular architecture in interfacial self-adhesion of polyethylene films*. Journal of Plastic Film & Sheeting, 2017. **33**(3): p. 235–261.

[7]   Liang, J.-Z., *Melt strength and drawability of HDPE, LDPE and HDPE/LDPE blends*. Polymer Testing, 2019. **73**: p. 433–438.

[8]   Bartels, C.R., et al., *Self-diffusion in branched polymer melts*. Macromolecules, 1986. **19**(3): p. 785–793.

[9]   Prasad, A., *A quantitative analysis of low density polyethylene and linear low density polyethylene blends by differential scanning calorimetery and fourier transform infrared spectroscopy methods*. Polymer Engineering & Science, 1998. **38**(10): p. 1716–1728.

[10]   Karian, H., *Handbook of Polypropylene and Polypropylene Composites, Revised and Expanded*. 2003, CRC press.

[11]   Bertin, D., et al., *Polypropylene degradation: Theoretical and experimental investigations*. Polymer Degradation and Stability, 2010. **95**(5): p. 782–791.

[12]   Karger-Kocsis, J., *Amorphous or Atactic Polypropylene, in Polypropylene: An A-Z Reference*, J. Karger-Kocsis, Editor. 1999, Springer Netherlands: Dordrecht. p. 7–12.

[13]   Yamada, K., et al., *Molecular orientation effect of heat-sealed PP film on peel strength and structure*. Advances in Materials Physics and Chemistry, 2015. **5**(11): p. 8.

[14]   Henderson, A.M., *Ethylene-vinyl acetate (EVA) copolymers: A general review*. IEEE Electrical Insulation Magazine, 1993. **9**(1): p. 30–38.

[15]   Wakabayashi, K. and R.A. Register, *Micromechanical interpretation of the modulus of ethylene–(meth)acrylic acid copolymers*. Polymer, 2005. **46**(20): p. 8838–8845.

[16]   Grady, B.P., *Review and critical analysis of the morphology of random ionomers across many length scales*. Polymer Engineering & Science, 2008. **48**(6): p. 1029–1051.

[17]   Tadano, K., et al., *Order-disorder transition of ionic clusters in ionomers*. Macromolecules, 1989. **22**(1): p. 226–233.

[18]   Farrell, K.V., *Effects of temperature on aggregate local structure in a zinc-neutralized carboxylate ionomer*. Macromolecules, 2000. **33**(19): p. 7122–7126.

[19]   Sakamoto, K., W.J. MacKnight, and R.S. Porter, *Dynamic and steady-shear melt rheology of and ethylene-methacrylic acid copolymer and its salts*. Journal of Polymer Science Part A-2: Polymer Physics, 1970. **8**(2): p. 277–287.

[20]   MacKnight, W., L. McKenna, and B. Read, *Properties of ethylene-methacrylic acid copolymers and their sodium salts: Mechanical relaxations*. Journal of Applied Physics, 1967. **38**(11): p. 4208–4212.

[21]   Morris, B.A. and J.M. Scherer, *Modeling and experimental analysis of squeeze flow of sealant during hot bar sealing and methods of preventing squeeze-out*. Journal of Plastic Film & Sheeting, 2015. **32**(1): p. 34–55.

[22]   Ma, Y., et al., *Solid-state polymerization of PET: Influence of nitrogen sweep and high vacuum*. Polymer, 2003. **44**(15): p. 4085–4096.

Chapter 5
Effect of processing and material properties on seal performance

Understanding the effects of different processing and material parameters on heat sealing is essential in controlling heat sealing properties of sealant films. In this chapter, a review of previous works that studied the effects of sealant material characteristics and heat sealing condition on seal and hot tack strength will be presented. The main sealing process parameters that affect seal performance are sealing temperature, dwell time, sealing pressure, and delay time or cooling time.

5.1 Sealing temperature

Sealing temperature is a critical parameter in heat sealing and needs to be adjusted carefully to ensure achieving the desired seal properties. Flexible packaging experiences a thermal shock once they are sandwiched between two heated jaws during heat sealing. For instance, if we assume that a film at room temperature is placed between two heated jaws at 150 °C and reaches jaw temperature after 0.5 s, the film experiences a heating rate of 300 °C/s or 18,000 °C/min. Such high heating rate during heat sealing process and the thin thickness of films make measuring temperature at the interface between seal sides very challenging. Therefore, in most previous studies, effects of jaw temperature on the seal performance were investigated instead of the effect of the interface temperature. Theller [1] examined heat sealing of cast coextruded low-density polyethylene (LDPE) films and blown coextruded high-density polyethylene (HDPE)/ethylene vinyl acetate (EVA)-PB films and showed that increasing jaw temperature increased the seal strength (SS). He also found that the highest peelable SS could be obtained around the melting temperature of the sealant material which is in agreement with the results of other researches [2]. Tetsuya et al. [3] showed a very low SS for oriented polypropylene (OPP)/cast polypropylene (CPP) laminate films sealed below ~110 °C. The maximum SS in this system achieved by increasing temperature to 120 °C but further increase in temperature reduced the SS. Similar results were reported by other researchers [4]. Morris [5] reported that high SS could only be achieved at temperatures higher than melting point of sealants. Najarzadeh and Ajji [6] also studied the effect of temperature on seal properties and concluded that a general optimum temperature cannot be defined as other process parameters and material characteristics also play a role in SS.

Muller and coworkers [7] observed a rapid increase in the SS by increasing the temperature from 115 to 125 °C in linear LDPE (LLDPE) sealant films. When they studied the morphology of the peeled surfaces at different temperatures, they found a

https://doi.org/10.1515/9781501524592-005

shift in the surface morphology from small, isolated fractured fibrils for the samples sealed below 115 °C to a membrane-like connected pattern in the samples sealed at 120 and 125 °C. They also found that increasing temperature up to 115 °C, increased the fibril density which was related to increasing chain diffusion across the interface. Increasing temperature above 115 °C led to a significant increase in the diffusion across the interface which increased the number and size of fibrils. Najarzadeh and Ajji [6] also studied the topography of the peeled surfaces of LLDPE films sealed at different temperatures using AFM and SEM and found that increasing sealing temperature increased the number of microfibrils or bridges between the surfaces which led to an increase in SS (Figure 5.1).

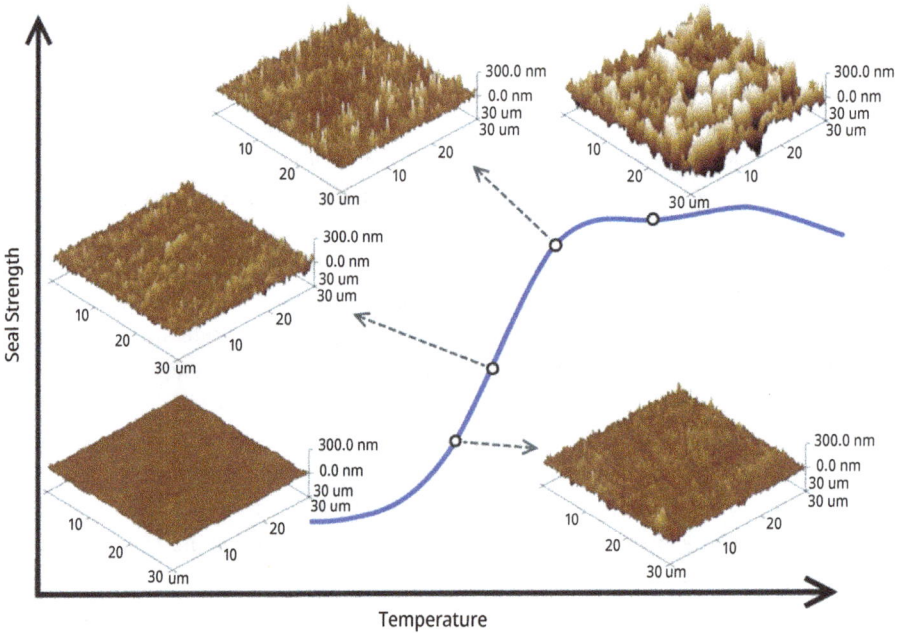

Figure 5.1: AFM images showing the effect of sealing temperature on the morphology of peeled surfaces that were sealed at dwell time= 0.5 s and pressure = 0.5 N/mm^2 [6].

These results point to the increase in the energy level of chain ends and the free volume that enhanced the chain end diffusion across the interface. When the temperature at the interface between seal sides reached the melting temperature of the sealant material, molecular diffusion across the interface was enhanced considerably. This increased the number and strength of the links created across the interfaces. When pulled apart, these links will be stretched into fibrils and finally break. Increasing temperature increases the number and strength of these links which increases the SS. Finally, the number of links increases so that it covers the majority of the surface and lock-seal behavior is observed. A similar trend can be seen in the previous studies on the effects of sealing temperature in which increasing sealing temperature increases the SS followed by reaching a plateau at higher temperatures and finally decay in SS at very high temperatures [2, 3, 6, 8–14]. Figure 5.2 shows an example of this behavior with the following important aspects:

- The seal initiation temperature (T_{si} or SIT): an onset temperature where the SS begins to increase rapidly with temperature.
- The plateau initiation temperature (T_{pi}): the temperature where the SS reaches a plateau and does not change considerably by further increase in temperature.
- Plateau SS (SS_p): the SS in the plateau region.
- The final plateau temperature (T_{pf}): the temperature in which the SS begins to decay rapidly, and extensive seal distortion is observed.

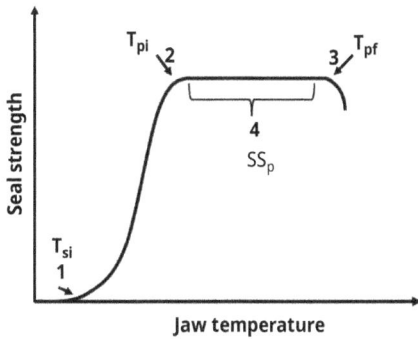

Figure 5.2: Example of seal strength variation with sealing temperature. The arrows show important points in the curve. T_{si}, seal initiation temperature; T_{pi}, seal plateau initiation temperature; SS_p, seal strength plateau; T_{pf}, seal strength decay [15, 16].

The definition of SIT depends on determining the SS threshold in which sealing begins. The value of this threshold varies between industries and different values from 100 to 500 g/in can be seen in literature. It should be noted that in some cases SIT is confused with T_{pi}. As a common role of thumb for comparing SIT of different materials, SIT follows the melting temperature of the sealant. For example in the case of polyethylene, the following order for SIT order is observed: very LDPE (vLDPE), LLDPE, and HDPE follow their melting temperature: SIT(vLDPE) < SIT(LLDPE) < SIT(HDPE), which matches with their melting temperatures.

T_{pi} indicates the temperature in which the seal reached its maximum strength. This temperature is important in determining sealing temperature where lock seal property is expected. The SS at plateau (SS_p) is important in designing the package where lock seal behavior is expected. In those applications, this value should be at least equal to the yield strength of the film to ensure that seal failure does not occur before film deformation. At T_{pf}, seal distortion and sealant squeeze out causes the decay in SS [6]. The sealing window or sealing range is defined as the temperature range between SIT and T_{pf}. From industrial viewpoint, a broad sealing window is an important parameter as it allows changing temperature to achieve optimum performance/cost balance or to compensate the variation of processing conditions such as temperature of jaws, humidity, or food contaminations [17].

The effect of sealing temperature on hot tack of sealants has also been studied in the literature and a general trend has been observed in which increasing temperature increases hot tack strength up to a maximum strength and further increase in temperature reduces it, as illustrated in Figure 5.3 [6, 18–27]. While increase in hot tack at temperatures below the maximum is attributed to the increase to the molecular diffusion, the observed decrease at higher temperature has been attributed to the easy pull out of polymer chains [14].

The different effect of temperature on seal and hot tack can be understood by considering the fact that seal test is done at room temperature where polymer is solid and therefore chain pull-out is much difficult. Therefore, failure mechanism in seal test is due to craze formation followed by stretching of fibrils. On the other hand, hot tack is done in the molten or hot polymer melt where chain pull-out requires much lower force. This makes chain pull-out the main failure mechanism in hot tack test.

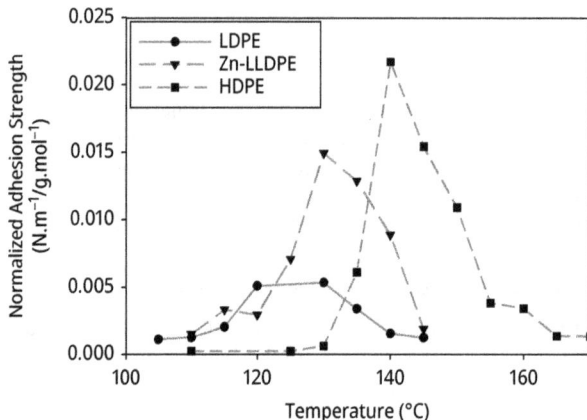

Figure 5.3: Examples of hot tack curved of LDPE, conventional LLDPE, and HDPE normalized to the molecular weight of the sealant [28].

5.2 Dwell time

During the dwell time, two phenomena should occur to achieve good seal or hot tack strength [1]: (i) the interface between seal layers should reach the temperature that allows chain ends to diffuse across the interface and (ii) Brownian motion of the chain ends across the interface should occur with a length scale large enough to create entanglement. Increasing dwell time increases the seal or hot tack strength by increasing both the interface temperature and diffusion depth of the polymer chain ends [27].

On the other hand, increasing dwell time above certain time does not improve considerably the SS as the interface temperature approaches the jaw temperature and the diffusion length becomes comparable to critical length scale required for entanglement [4]. Therefore, an optimum dwell time that ensures strong seal while allowing high production rate needs to be determined for any desired application. It has been well established in literature that increasing the sealing temperature reduces the required dwell time to reach high SS through increasing heating rate and improved molecular diffusion [3, 4, 6, 15, 16, 29].

Yuan et al. [4] showed that in two layers laminated films of OPP/CPP, increasing dwell time at temperatures below seal initiation temperature (T_{si} = 122 °C) did not have a considerable effect but above this temperature, the effect of dwell time was considerable. They also found that the effect of dwell time became negligible (even at temperatures above T_{si}) after reaching plateau SS. Based on the reptation model, Mueller et al. [7] proposed a linear relation between SS and square root of dwell time. This relation was confirmed later by Qureshi et al. [30] and Najarzadeh and Ajji [6].

Najarzadeh and Ajji [6] plotted a 3D diagram (Figure 5.4) to show the combination effects of dwell time and jaw temperature on SS and found a strong dependency of SS on both parameters. They found that there is an optimum region of dwell time and temperature which results in the highest SS (the orange regions in Figure 5.4).

Dwell time has been shown to have a similar effect on hot tack, due to its effects on molecular diffusion of chains across the interface [14, 31, 32]. This indicates that the combination of sealing temperature and dwell time provides a powerful design tool to control and optimize seal properties. Generally, in flexible packaging industry, dwell time is of the order of fractions of a second, or in some cases 1–2 s for thick samples. The optimum dwell time ensures that no excessive time is lost to keep up the production rate.

5.3 Sealing pressure

Sealing pressure is needed to bring seal layers to close contact and overcome microroughness between surfaces to enhance wetting and eventually molecular diffusion across the interface [1]. For example, Theller [1] observed that increasing pressure from 0.27 to 2.75 MPa changed slightly the SS of HDPE films but did not have a considerable effect on the SS of LDPE films. Accordingly, the effects of dwell time and sealing temperature are much important than sealing pressure. This conclusion has also been reported by many researchers [4, 6, 9, 15, 16, 33]. It has been mentioned that excessive pressure may squeeze out the molten fraction of material from the seal area and distort the seal [6]. A minimum required pressure has also been reported which below that, no SS could be attained. It should be mentioned that this minimum pressure limit depends on the material, film thickness, and surface roughness of the films and a general limit cannot be presented [33].

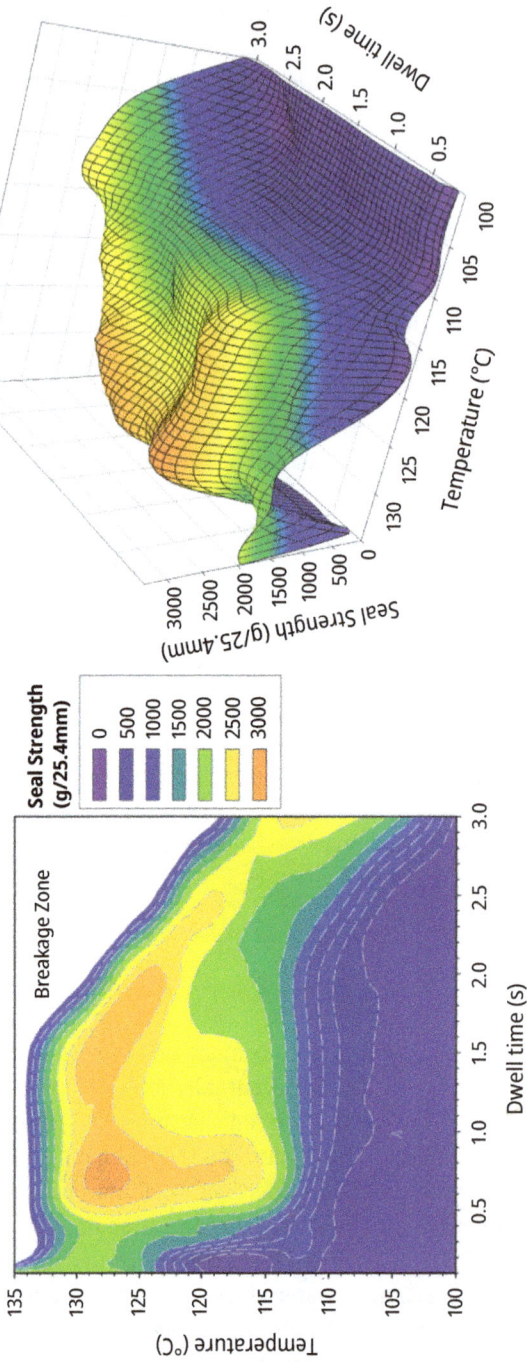

Figure 5.4: 2D (left) and 3D (right) plot of seal strength versus different temperatures and dwell times [6].

Najarzadeh and Ajji [6] presented a 3D diagram that shows the relation between pressure, temperature, and SS of metallocene PE films. The orange region in Figure 5.5 shows the optimum pressure and temperature that resulted in the highest SS. They found that a minimum pressure of 0.2 MPa at temperatures below 125 °C was needed to achieve good SS while at higher sealing temperatures, this limit disappeared. They also found that increasing pressure to a high value of 3 MPa allowed a reduction in the temperature range in which they could observe the high SS. The observed reduction in SS at high seal pressure was attributed to the squeeze out flow of the film at high pressures. Similar effect of sealing pressure has been reported for hot tack [31, 34–36] with a minimum pressure required to ensure the contact between the films during heat sealing.

Some defects such as channeling or wrinkles due to contamination and three-point junctions threaten the sealing functionality. Their effect can also be minimized by selecting sealants with good caulkability [37–39]. Caulkability is the ability of a sealant material to flow and fill gaps in the wrinkles and junctions or encapsulate contaminations in the seal area. This property is especially needed for packaging of powder grainy or shredded products such as coffee powder or shredded/powdered cheese. Considering the physics of heat sealing process, caulkability can be seen as the ability of a sealant to flow under the squeezing sealing pressure during heat sealing process. Therefore, selecting the optimum sealing pressure beside the good temperature and dwell time could help to overcome these issues [40]. It should be noted that caulkability also depends on material properties that should be considered as well.

5.4 Effect of material characteristics

5.4.1 Effect of crystallinity

During the heating stage of heat sealing cycle, polymer crystals have to be melted to allow molecular diffusion across the interface. In addition, after opening the heated jaws at the end of dwell time, the sealed area begins to cool down due to the heat transfer to the surrounding environment. Cooling of the sealed area in semicrystalline polymers can be considered as a nonisothermal crystallization process. In semicrystalline polymers, crystallization can occur mainly between melt temperature (T_m) and glass transition temperature (T_g) [41].

Crystallinity of sealants affects both seal and hot tack properties as increasing their crystallinity increases their melting temperature and retards chain diffusion. After sealing, recrystallization of diffused chains during delay time increases seal and hot tack strength. For example, Stehling and Meka [16] studied sealing behavior of 42 different types of sealants including m-LLDPE, ZN-LLDPE, HDPE, LDPE, VLDPE, PP, and

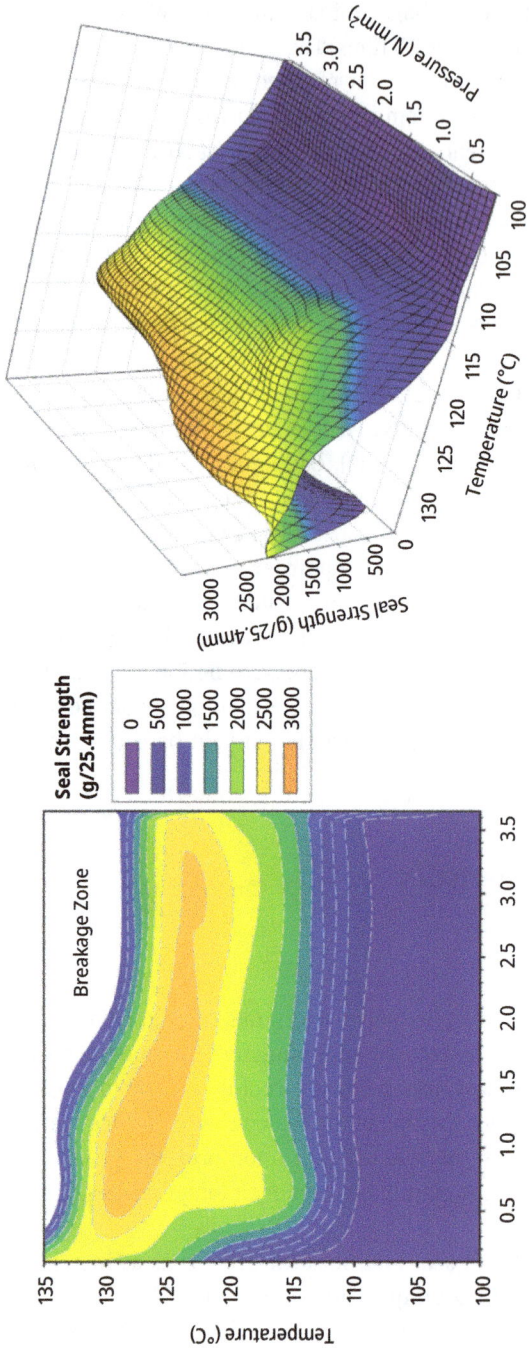

Figure 5.5: 2D countour (left) and 3D plot (right) of seal strength at different sealing pressure and temperature for metallocene PE films. Sealing was done at dwell time of 0.5 s [6].

EVA. They estimated the amorphous fraction at any temperature, f_a (T), using dynamic scanning calorimetry (DSC) and the following equation:

$$f_a(T) = 1 - \left[\frac{\Delta H_S - \Delta H_T}{\Delta H_U}\right] \tag{5.1}$$

where ΔH_T, ΔH_S, and ΔH_U are the cumulative heat of fusion at temperature T, the total heat of fusion of the sample, and the heat of fusion of 100% crystalline polymer, which is 293 J/g for polyethylene [42]. They found that for semicrystalline polymers, amorphous fraction at sealing temperature has a significant effect on SS and concluded that seal initiation temperature occurs when amorphous fraction reaches 77%. In addition, a broader distribution of f_a as a function of temperature results in broader transition region between seal initiation temperature and seal plateau. Similar results and conclusions were reported by Najarzadeh and Ajji [6]. Stehling and Meka [16] concluded that the shape of the seal curve between SIT and seal plateau depends on distribution of the melting peak.

The addition of high molecular weight (HMW) fractions to increase amorphous fraction of sealant materials has been examined by some researchers as a method to improve seal properties of highly crystalline polymers. For example, Miyata et al. [43] compared the wide-angle X-ray scattering pattern and DSC scans of HDPE blends with different amounts of HMW HDPE before and after heat sealing. Their results showed that the addition of less than 10% of HMW HDPE reduced crystallinity and crystalline orientation and enhanced SS. Some researchers also showed that the addition of low crystalline ethylene-based copolymers such as EVA or plastomers reduced seal initiation temperature and increased SS by reducing the crystallinity of polyethylene matrices [44–47]. Moreira et al. [27] attempted to establish a relation between Avrami crystallization rate and seal properties and concluded that good seal properties could be obtained when the Avrami crystallization rate at the sealing temperature was greater than 0.2. Nicastro et al. [48] studied the crystallinity of PP during heat sealing and its effect on SS. By comparing the crystallinity of films before and after heat sealing, they found that crystallinity of PP films increased from 38% for the film before heat sealing to 50–56% after cooling of the sealed area. By comparing SS of film samples with different crystallinity, they claimed that they found a linear relation between SS and crystallinity of the film, although the linear behavior is not clear from their results.

The observed different effect of crystallinity in the previous studies can be related to the complex effect of crystallinity on the seal properties of polymer films. Low crystallinity level enhances melting and diffusion during sealing while high crystallinity can provide higher SS after cooling of the seal.

5.4.2 Effect of molecular architecture

Previous studies showed that increasing molecular weight (MW) of the sealant material increased the plateau SS, but also increased the seal initiation temperature (T_{si}) [7, 14, 16, 24, 44]. Najarzadeh [14] compared seal properties of three metallocene LLDPE (m-LLDPE) with different MWs, narrow molecular weight distribution (MWD), and almost identical branch densities. She found that increasing MW increased both hot tack strength and seal initiation temperature (Figure 5.6, left). She also reported a linear relation between the maximum hot tack strength and MW (Figure 5.6, right). They observed that the effect of MW on seal properties can be attributed to higher entanglements of polymer chains by increasing their MW [49].

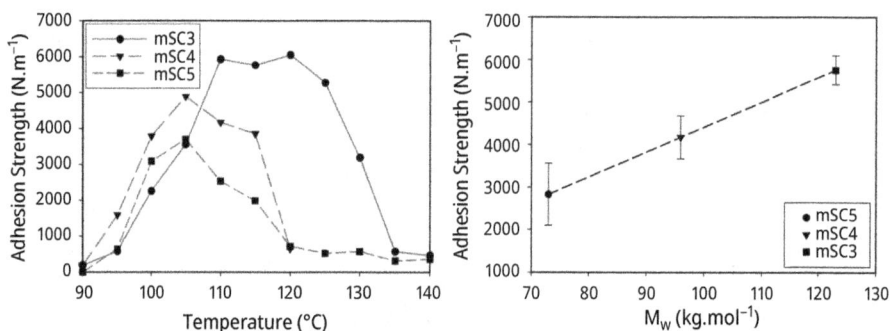

Figure 5.6: (Left) The effect of molecular weight on hot tack or adhesion strength of metallocene short chain branched linear low-density polyethylene. (Right) Linear relationship between maximum hot tack strength and MW. MW of mSC3, mSC4, and mSC5 are 123, 96, and 73 kg/mol, respectively [14].

5.4.3 Chain branching

There are two types of chain branching: long chain branches (LCB) and short chain branches (SCB). If MW of branches is greater than critical entanglement MW (M_e) (which is for example around 2,800 g/mol for PE or 100 c–c bond length [41]), the branch is considered as an LCB [6]. Branches with lower MW than this critical value are considered as SCB. Najarzadeh and Ajji [14, 46] studied the effect of branching on hot tack of a conventional polyethylene with broad MWD and two metallocene long-chain branched ethylene α-olefin copolymers with narrow MWD.

Table 5.1 summarizes the molecular characteristics of the materials studied in their work. Their hot tack results are also shown in Figure 5.7.

Table 5.1: Molecular characteristics of branched PEs studied by Najarzadeh and Ajji [28].

Type	Code	Mw (kg/mol)	MWD	LCB density (1/10^4 C atoms)	SCB density (1/10^4 C atoms)
Conventional LDPE	LDPE	160	8.75	5.2	–
Metallocene LLDPE	mLC1	115	2.6	0.19	2.4
Metallocene LLDPE	mLC2	115	2.1	0.3	3.7

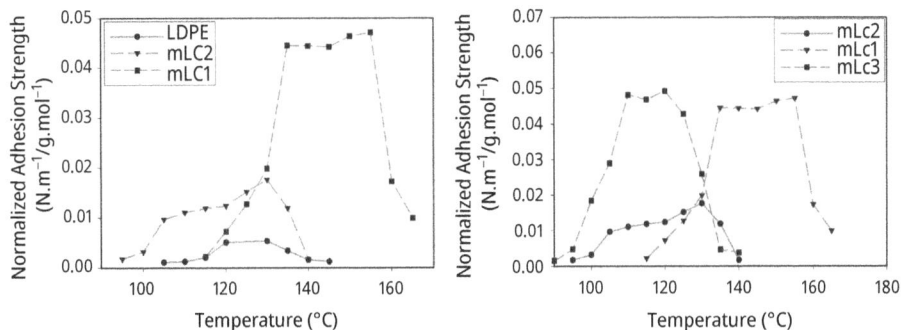

Figure 5.7: *Left*: Adhesion strength or hot tack normalized by M_w versus seal temperature for a conventional LDPE and two metallocene LLDPE. *Right*: Comparing normalized adhesion strength of LCB metallocene LLDPEs with SCB metallocene LLDPEs (mLC3 had an M_w of 123 kg/mol and SCB density of 2.5 (1/10,000 C atom)) [14].

As shown in Figure 5.7 (left), increasing the density of LCBs from 0.19 to 0.3 (per 10^4 C atoms) reduced significantly hot tack strength. Further increase in LCB density to 5.2 (in conventional LDPE) dramatically reduced hot tack properties. These results indicate that the presence of LCB increases seal initiation temperature and reduces significantly hot tack strength. Similar effects of LCB on hot tack were reported by Moreira et al. [27] for metallocene polyethylene. The effect of LCB on seal and hot tack strength can be attributed to much difficult chain diffusion in the presence of LCB due to the increased entanglement density with the surrounding chains [50]. As shown in Table 5.1, the SCB branching densities of mLC1 and mLC2 are different; therefore, Najarzadeh and Ajji [28] also compared their hot tack results with the hot tack results of a short-chain branched polyethylene and their results are presented in Figure 5.7 (right). As the SCB densities of mLC3 and mLC1 are almost identical, comparing the results of these two samples can clearly show that LCB retarded the diffusion due to an increased hindrance of LCB on molecular motion.

The observed differences between seal performance of metallocene PE and conventional PE have motivated some researchers to investigate the origin of these differences. Previous researchers reported surface segregation in conventional polyethylene films in which an amorphous layer composed of highly branched and low MW chains exist at the surface [30, 51–54]. The observed segregation at the surface is attributed to the entropic gain due to the presence of low MW chains at the surface. Moreover, the presence of highly branched copolymers increases significantly the chain end density at the surface which reduces the surface energy by increasing surface entropy [30]. During heat sealing of conventional PE films, the low MW fraction in the segregated layer can diffuse rapidly but they cannot provide SS due to the lack of entanglement formation. The highly branched fraction of the segregated layer requires long dwell times to diffuse. Both phenomena believed to cause the poor seal properties observed for conventional PE films. On the other hand, the much narrower MW in metallocene PE leads to the formation of much uniform chain size and branching distribution through the film. These results clearly highlight the importance of high degree of control of MW and branching density distribution in achieving high seal properties.

5.4.4 Monomer sequence

In addition to the effect of branching and MW on seal performance, few works studied the effect of monomer sequences in ethylene alpha olefin copolymers. Moreira et al. [27] tried to find a relation between ethylene sequence in LLDPE and seal properties by defining a new parameter which they called the welding power. They ran hot tack tests at dwell times between 0.2 and 2 s and then plotted force/25 mm as a function of the dwell time. The welding power was then defined as the initial slope of this curve at dwell times between 0.2 and 0.5 s. They reported that increasing the ethylene dispersity (which means shorter ethylene sequences in LLDPE chains) decreased the seal initiation temperature by reducing the melting temperature.

References

[1]   Theller, H.W., *Heat sealability of flexible web materials in hot-bar sealing applications*. Journal of Plastic Film and Sheeting, 1989. **5**: p. 66–93.

[2]   Aithani, D., et al., *Heat Sealing measurement by an innovative technique*. Packaging Technology and Science, 2006. **19**(5): p. 245–257.

[3]   Tetsuya, T., et al., *The effect of heat sealing temperature on the properties of OPP/CPP heat seal. I. Mechanical properties*. Journal of Applied Polymer Science, 2005. **97**(3): p. 753–760.

[4]   Yuan, C. and H. A, *Effect of bar sealing parameters on OPP/MCPP heat seal strength*. Express Polymer Letters, 2007. **1**(11): p. 773–779.

[5]   Morris, B.A., *Predicting the heat seal performance of ionomer films*. Plastic Film and Sheeting, 2002. **18**.

[6]   Najarzadeh, Z. and A. Ajji, *A novel approach toward the effect of seal process parameters on final seal strength and microstructure of LLDPE*. Journal of Adhesion Science and Technology, 2014. **28**(16): p. 1592–1609.

[7]   Mueller, C., et al., *Heat sealing of LLDPE: Relationships to melting and interdiffusion*. Journal of Applied Polymer Science, 1998. **70**(10): p. 2021–2030.

[8]   Hashimoto, Y., et al., *Effect of Heat-sealing Temperature and Holding Time on Mechanical Properties of Heat-sealed Poly(Lactic Acid) Films*. 2006, ANTEC.

[9]   Poisson, C., et al., *Optimization of PE/Binder/PA extrusion blow-molded films. I. Heat sealing ability improvement using PE/EVA blends.* Journal of Applied Polymer Science, 2006. **99**(3): p. 974–985.

[10]   Yuan, C.S., et al., *Heat sealability of laminated films with LLDPE and LDPE as the sealant materials in bar sealing application.* Journal of Applied Polymer Science, 2007. **104**(6): p. 3736–3745.

[11]   Planes, E., S. Marouani, and L. Flandin, *Optimizing the heat sealing parameters of multilayers polymeric films.* Journal of Materials Science, 2011. **46**(18): p. 5948–5958.

[12]   Mazzola, N., et al., *Correlation between thermal behavior of a sealant and heat sealing of polyolefin films.* Journal of Polymer Testing, 2012. **31**(7): p. 870–875.

[13]   Robertson, G.L., *Food Packaging, Principles and Practice.* Third ed. 2012, CRC Press.

[14]   Najarzadeh, Z., Control and optimization of sealing layer in films, PhD Thesis, École Polytechnique de Montréal, 2014.

[15]   Meka, P. and F.C. Stehling, *Heat sealing of semicrystalline polymer films. I. Calculation and measurement of interfacial temperatures: Effect of process variables on seal properties.* Journal of Applied Polymer Science, 1994. **51**(1): p. 89–103.

[16]   Stehling, F.C. and P. Meka, *Heat sealing of semicrystalline polymer films. II. Effect of melting distribution on heat-sealing behavior of polyolefins.* Journal of Applied Polymer Science, 1994. **51**(1): p. 105–119.

[17]   Pellingra, S., *Improving Line Efficiencies with Sealant Optimization*, AmpacPackaging, Editor. 2009.

[18]   Kanani Aghkand, Z., A. Saffar, and A. Ajji, *Effect of Back-layer on seal performance of multilayer polyethylene-based sealant films.* Journal of Applied Polymer Science, 2021. **138**(31): p. 50742.

[19]   Brodie. *Hot tack procedures for flexible packaging structures.*

[20]   Vincent, B. *Hot tack of sealant resins.* in *TAPPI, Polymers, Laminations and coatings conference.* 1986.

[21]   Kiang, W., *Hot tack and heat sealing properties of EVOH,* in *TAPPI place, lamination and coating.* 1992.

[22]   Philippe Mesnil, et al., *Seal through contamination performance of metallocene plastomers.* In *TAPPI Polymers, Laminations & Coatings Conference.* 2000, Chicago.

[23]   Halle, R.W. and K.M. Cable, *A New mLLDPE for Extrusion Coating Applications.* ExxonMobil Chemical Company, 1993.

[24]   Halle, R.W. *Plastomer-mVLDPE Blends for High Performance Heat Sealing Applications.* In *TAPPI 2003 PLACE Conference Proceedings.* 2003.

[25]   Shih, H.-H., et al., *Hot tack of metallocene catalyzed polyethylene and low-density polyethylene blend.* Journal of Applied Polymer Science, 1999. **73**(9): p. 1769–1773.

[26]   Shekhar, A., *A model for hot tack behavior in Ethylene Acid co-polymer films.* Tappi, 1994. **77**(1): p. 97–104.

[27]   Moreira, A.C.F., P.C. Dartora, and F. Paulo dos Santos, *Polyethylenes in blown films: Effect of molecular structure on sealability and crystallization kinetics.* Polymer Engineering & Science, 2017. **57**(1): p. 52–59.

[28]   Najarzadeh, Z. and A. Ajji, *Role of molecular architecture in interfacial self-adhesion of polyethylene films.* Journal of Plastic Film & Sheeting, 2017. **33**(3): p. 235–261.

[29]   Kim, S., et al., *Enhanced interfacial adhesion between an amorphous polymer (Polystyrene) and a semicrystalline polymer [a Polyamide (Nylon 6)].* ACS Applied Materials & Interfaces, 2011. **3**(7): p. 2622–2629.

[30]   N. Z. Qureshi, et al., *Self-adhesion of polyethylene in the Melt. 1. Heterogeneous Copolymers.* Macromolecules, 2001. **34**: p. 1358–1364.

[31]   Ebnesajjad, S., *Plastic Films in Food Packaging: Materials, Technology and Applications.* 2012, William Andrew.

[32]   Theller, H.W., *Heat sealability of flexible web materials in hot-bar sealing applications.* Plastic Film and Sheeting, 1989. **5**: p. 66–93.

[33]   Selke, S.E.M., J.D. Culter, and R.J. Hernandez, *Plastics Packaging: Properties, Processing, Applications, and Regulations*, ed. 2nd. 2004, Hanser Gardner Publications.

[34] Sierra, J.D., M. Del Pilar Noriega, and T.A. Osswald, *Effect of metallocene polyethylene on heat sealing properties of low density polyethylene blends*. Journal of Plastic Film & Sheeting, 2000. **16**(1): p. 33–42.

[35] Kiang, W., *Hot tack and heat sealing properties of EVOH*. ANTEC 92 – Shaping the Future, 1992. 1: p. 1244–1249.

[36] Morris, B.A., *The Science and Technology of Flexible Packaging: Multilayer Films from Resin and Process to End Use*. 2016, William Andrew.

[37] Ward, D. and M. Li, *Seal-through-contamination and "caulkability" an evaluation of sealants' ability to encapsulate contaminants in the seal area*, In *TAPPI PLACE Conference*. 2016, Fort Worth, Texas.

[38] Morris, B.A. and J.M. Scherer, *Modeling and experimental analysis of squeeze flow of sealant during hot bar sealing and methods of preventing squeeze-out*. Journal of Plastic Film and Sheeting, 2015. **32**(1): p. 34–55.

[39] Sadeghi, F. and A. Ajji, *Application of single site catalyst metallocene polyethylenes in extruded films: Effect of molecular structure on sealability, flexural cracking and mechanical properties*. The Canadian Journal of Chemical Engineering, 2014. **92**(7): p. 1181–1188.

[40] Kanani Aghkand, Z. and A. Ajji, *Squeeze flow in multilayer polymeric films: Effect of material characteristics and process conditions*. Journal of Applied Polymer Science, 2022: p. 51852.

[41] Sperling, L.H., *Introduction to Physical Polymer Science*. 4th ed. 2006, John Wiley & Sons, Inc.: New Jersey.

[42] Zhu, L., et al., *Physical Constants of Poly(ethylene), in Polymer Handbook*. 4th ed., J. Brandrup, et al., Editors. John Wiley & Sons. p. V-9.

[43] Miyata, K., et al., *The relationships between crystallization characteristics and heat sealing properties of high-density polyethylene films*. Journal of Plastic Film and Sheeting, 2014. **30**(1): p. 28–47.

[44] Morris, B.A., *Predicting the performance of ionomer films in heat-seal processes*. Paper, Film and Foil Converter, 2003. **77**(3): p. 207.

[45] Poisson, C., et al., *Optimization of PE/binder/PA extrusion blow-molded films. II. Adhesion properties improvement using binder/EVA blends*. Journal of Applied Polymer Science, 2006. **101**(1): p. 118–127.

[46] Najarzadeh, Z., A. Ajji, and J.-B. Bruchet, *Interfacial self-adhesion of polyethylene blends: The role of long chain branching and extensional rheology*. Rheologica Acta, 2015. **54**(5): p. 377–389.

[47] Chen, Y., et al., *Melting and crystallization behavior of partially miscible high density polyethylene/ ethylene vinyl acetate copolymer (HDPE/EVA) blends*. Thermochimica Acta, 2014. **586**: p. 1–8.

[48] L. C. Nicastro, et al., Change in crystallinity during heat sealing of cast polypropylene film. *Plastic Film and Sheeting*, 1993. **9**.

[49] Bartels, C.R., B. Crist, and W.W. Graessley, *Self-diffusion coefficient in melts of linear polymers: Chain length and temperature dependence for hydrogenated polybutadiene*. Macromolecules, 1984. **17**(12): p. 2702–2708.

[50] Bartels, C.R., et al., *Self-diffusion in branched polymer melts*. Macromolecules, 1986. **19**(3): p. 785–793.

[51] Qureshi, N.Z., et al., *Self-adhesion of polyethylene in the Melt. 2. Comparison of heterogeneous and homogeneous copolymers*. Macromolecules, 2001. **34**(9): p. 3007–3017.

[52] David T. Wu and G.H. Fredrickson, *Effect of architecture in the surface segregation of polymer blends*. Macromolecules, 1996. **29**: p. 7919–7930.

[53] Schuman, T., et al., *Solid state structure and melting behavior of interdiffused polyethylenes in microlayers*. Polymer, 1999. **40**(26): p. 7373–7385.

[54] Schuman, T., et al., *Interdiffusion of linear and branched polyethylene in microlayers studied via melting behavior*. Macromolecules, 1998. **31**(14): p. 4551–4561.

Chapter 6
Modeling of heat sealing process

Optimization of heat sealing process parameters to obtain a desired seal strength at shorter dwell times and lower jaw temperatures is an approach for increasing production rate and reducing final product cost. Optimization of heat sealing lines in packaging industry is commonly done using experimental trial and error which requires considerable amount of time, material, and energy. As a result, heat sealing machines are usually run at much higher temperatures than melting temperatures of sealant layers to ensure sealing at short dwell times. In order to show the importance of optimization in an industrial heat sealing process, a heat sealing machine with line speed of 100 packages per minute and dwell time of 0.4 s can seal more than 52 million packages per year. The production capacity of this line can be increased to more than 63 million packages by only reducing 0.1 s of the dwell time. This will also increase the profit and reduce the final cost and result in a more competitive final price for the product. It should be noted that reducing the dwell time reduces the power consumption of the heaters used in the jaws as well.

Considering the mentioned drawbacks and limitations of experimental optimization technique of the heat sealing process, optimization based on mathematical modeling and simulation software has been introduced as an promising strategy to reduce optimization time and cost [1, 2]. A comprehensive model of heat sealing process should consider (i) heat transfer from the jaws toward the interface between two seal sides, (ii) molecular interdiffusion at the interface, (iii) squeeze-out flow, (iv) crystallization after sealing, and (v) estimation of seal strength development. Due to the complexity of presenting a model with all components, a comprehensive predictive heat sealing model is lacking in the literature. As shown in Chapter 5, interface temperature has a critical role in determining the final seal performance. Therefore, first, previous studies on modeling or simulation of the interface temperature in the heat sealing process will be reviewed in this chapter. It should be noted that theories of molecular interdiffusion and crystallization from the melt were thoroughly discussed in Chapter 2 and, therefore, will not be reviewed here. Previous studies attempted to model the squeeze-out flow will also be reviewed later in this chapter. Finally, different mathematical models proposed for estimation of seal strength development during heat sealing will be reviewed at the end of this chapter.

6.1 Modeling of the interface temperature

Modeling of the interface temperature using material characteristics and processing condition is an interesting approach to predict the interface temperature in a heat sealing process. Figure 6.1 shows the configuration of a multilayer film between

https://doi.org/10.1515/9781501524592-006

heated jaws in a heat sealing process. In most cases, heated jaws are at the same temperature (symmetric heating) and heat transfer is from the heated jaws toward the interface. In some cases, only one jaw is heated (asymmetric heating) and the heat transfer in these cases is from the heated jaw toward the cold jaw.

Kanani Aghkand et al. [3] measured interface temperature variations in different directions in the sealed area. Their results are shown in Figure 6.2. They found that only the temperature gradient in the thickness direction is considerable. Therefore, the heat sealing process can be considered as a one-dimensional heat transfer problem in the thickness direction.

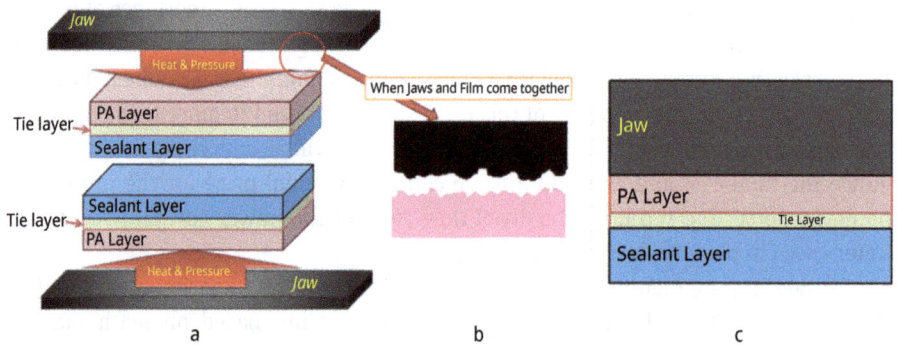

Figure 6.1: (a) Geometry of a multilayer film in contact with a heated jaw in a heat sealing process and (b) microstructure of the contact area between jaw and the outermost layer of the film [3].

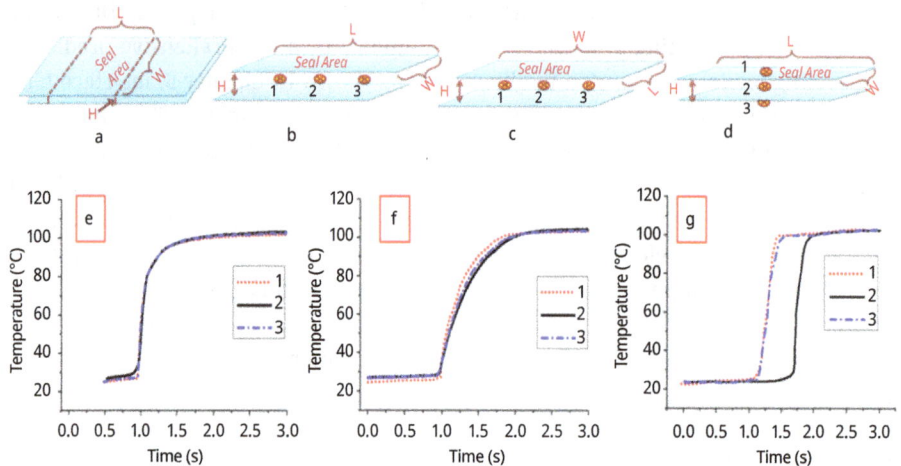

Figure 6.2: Variations of temperature in different directions in the sealed area: (a) seal area dimensions; (b) location of thermocouples in longitudinal direction; (c) width direction; and (d) thickness direction; (e), (f), and (g) show the recorded temperature profiles for thermocouples in (b), (c), and (d), respectively [3].

By considering constant thermal conductivity (k), density (ρ), and specific heat (C_p), one-dimensional transient heat transfer equation can be written as

$$\rho C_p \frac{dT}{dt} = -k\frac{d^2 T}{dx^2} - Q \tag{6.1}$$

where T, t, x, and Q are temperature, time, coordinate in thickness direction, and the heat absorbed by melting of polymer(s). Figure 6.3 shows schematically the solution of equation (6.1) in two cases of symmetric heating and asymmetric heating.

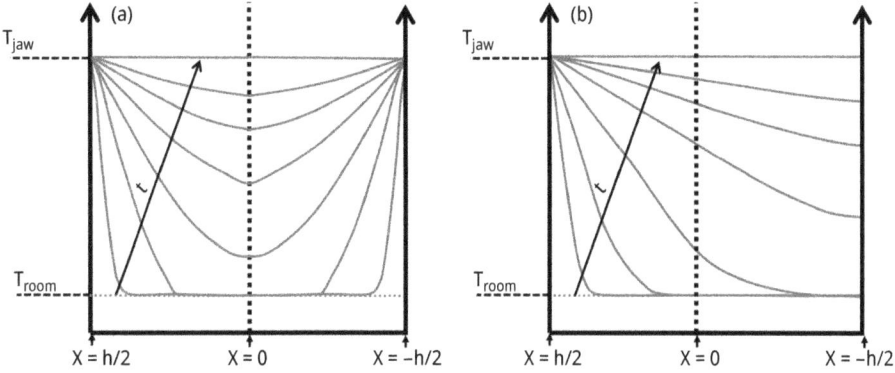

Figure 6.3: Schematics of temperature profile (gray lines) within two single-layer films with thickness of $h/2$ in a heat sealing process: (a) both jaws are heated (symmetric heating) and (b) only one jaw is heated (asymmetric heating) and the other jaw is assumed as an insulator.

Based on equation (6.1), heat absorbed during melting of sealant or other layers shifts the temperature profiles shown in Figure 6.3 toward lower temperatures and results in longer time required for the interface temperature to reach the jaw temperature. In order to model heat transfer in a heat sealing process, heat transfer at contact surface between two seal sides and heat transfer within the bulk of the films need to be discussed.

6.1.1 Heat transfer in contact areas

Heat transfer between surfaces at contact points has been the subject of many studies [4–8]. The main surface contact areas in a heat sealing process are (i) the contact between jaws and the films and (ii) the contact between two sides of the seal before their melting. The second contact area will be important only when asymmetric heating is applied, and in the case of symmetric heating, the surface contact between sides of the seal does not play a considerable role on heat transfer. This is due to the fact that in symmetric heating, heat transfer occurs from both jaws toward the interface between films and therefore the interface between two film is located at the symmetry boundary condition line and does not affect the heat transfer. As symmetric heating is used in most heat sealing processes especially in flexible packaging, the main focus in this part of the chapter is dedicated to the heat transfer from jaw to the film. It should be noted that the same concept can be used to model polymer/polymer contact between two sides of the seal in asymmetric heating sealing. Meka and Stehling [9] considered heat convection as the heat transfer mechanism from the heated jaws to the film:

$$Q = h\left(T_{\text{jaw}} - T_{\text{film}}\right) \tag{6.2}$$

where Q and h are the heat transferred from the jaw and the heat transfer coefficient (h), respectively. The authors used h as a fitting parameter to fit their model on experimental data of interface temperature. Their results showed that changing the heat transfer coefficient significantly affected simulation predictions. They found the best agreement between experimental and simulation results by using a heat transfer coefficient of 3,910 W/m^2 K. Mihindukulasuriya and Lim [10] used COMSOL and examined asymmetric heat sealing of two LLDPE films. They considered a heat transfer coefficient between coated jaw and the film and between two sides of the film. Similar to Meka and Stehling [9], they determined the heat transfer coefficients by fitting their model on the experimental results. The main disadvantage of finding heat transfer coefficients by fitting the model on experimental results is the fact that the obtained model cannot be used as a predictive model for other cases. In addition, using this approach may mask other phenomena occurring in the film such as variation of material properties and introduces error in the model results.

As heat sealing is usually done below melting temperature of the abuse layer, this layer remains in solid state during the heat sealing process. Figure 6.1(b) shows a microscale schematic of the contact area between the surface of the jaw and the abuse layer. The presence of microroughness on the film surface and the jaw surface prevents a complete contact between two surfaces. The presence of air pockets trapped between surfaces results in heat transfer through two simultaneous mechanisms: (i) heat conduction in the direct contact points and (ii) heat transfer through trapped air pockets. The resistance against heat transfer due to the presence of surface roughness is known as thermal contact resistance (TCR) and is considered in heat transfer models using an overall heat transfer coefficient (h). Heat transfer at the contact of two surfaces has been extensively studied in thermal contact theory. Based on this theory, the overall heat transfer coefficient (h) is defined as the sum of the gap conductance (h_g) and the heat transfer coefficient between contact surfaces (h_c). It should be noted that h_g and h_c indicate the contributions of heat transfer through trapped air and the contribution of heat transfer in the direct contact points, respectively. Gap conductance is defined as [6, 8]

$$h_g = k_g/\delta \tag{6.3}$$

where k_g is the gas conductivity and δ is the effective gap thickness which is related to the surface roughness, microhardness, and contact pressure. Different models for prediction of the effective gap thickness were reviewed by Song et al. [11].

Existing models for estimation of h_c can be categorized into elastic, plastic, and elastoplastic models based on the type of deformation of surface asperities [4–7, 12]. When the applied contact pressure by the jaws is greater than the microhardness of the softer surface, asperities on the softer surface undergo a plastic deformation. On the other hand, when the contact pressure is lower than the surface microhardness, only an elastic deformation occurs in surface asperities. Elastoplastic models consider a combination of the plastic and elastic deformations for surface asperities. Determining type of the deformation of surface asperities is important in estimation of h_c and, consequently, TCR as it directly affects the direct contact area between surfaces.

As heat sealing process is commonly done at temperatures above the glass transition temperature (T_g) of the commonly used abuse layers (PET or PA), Kanani Aghkand et al. [3, 13] recommended to use plastic Cooper–Mikic–Yovanovich (CMY) model in simulation of heat sealing process. CMY model uses the following equation for prediction of h_c:

$$h_c = 1.25 K_{\text{contact}} \frac{m_{\text{asp}}}{\sigma_{\text{asp}}} \left(\frac{P}{H_c} \right)^{0.95} \tag{6.4}$$

where P, H_c, m_{asp}, and σ_{asp} are the contact pressure, microhardness of the abuse layer surface, surface roughness (σ_{asp}), and asperity average slope (m_{asp}). In addition, the thermal conductivity between two contact surfaces (K_{contact}) is defined as

$$K_{\text{contact}} = \frac{2K_1 K_2}{K_1 + K_2} \tag{6.5}$$

where K_1 and K_2 are conductivities of the two surfaces. Kanani Aghkand et al. [3, 13] used COMSOL multiphysics software to model symmetric heat sealing process of a PA/tie/LLDPE film sample between two stainless steel jaws. They estimated TCR by CMY model using the measured surface microhardness, surface roughness, and thermal conductivity of the abuse layer. Figure 6.4 compares their simulation results and experimental data with and without TCR boundary condition.

The simulation results without TCR boundary condition are much different from the experimental result which emphasizes on the important role of considering TCR boundary condition in prediction of the interface temperature at early stages of the process (short dwell times). As most industrial heat sealing machines work at short dwell times (0.3–0.5 s), these results indicate the critical role of considering TCR in the simulation of heat sealing.

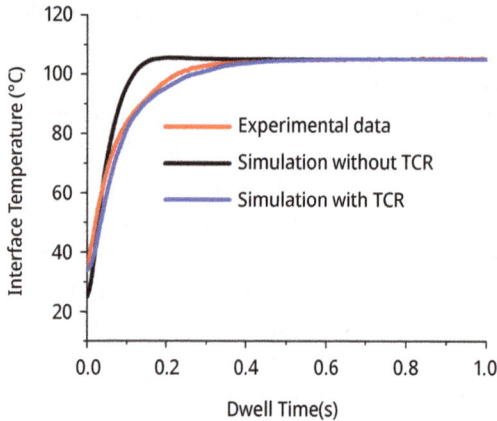

Figure 6.4: Comparing experimental data for the interface temperature and the simulation results with and without TCR boundary condition (experimental data from Kanani Aghkand et al. [3]).

6.1.2 Heat transfer within the film

Heat transfer within film structure can be described by the Fourier's law of heat conduction:

$$q = -kT \qquad (6.6)$$

where q is the heat flux, k is the thermal conductivity, and T is the temperature gradient. Meka and Stehling [9] modeled heat transfer in a single-layer sealant using finite element method (FEM) and compared the model predictions with experimental measurements of the interface temperature. They assumed isotropic and constant thermal conductivity, constant specific heat, and variable density in their model. In order to consider heat absorbed during melting of polymers, they also considered heat of fusion in some samples. Based on the comparison of their model predictions and experimental data, they found that their model predictions were satisfactory below melting temperature of the sealant layer. However, at temperatures above melting temperature of the sealant, model predictions were always lower than the measured interface temperatures. The authors attributed these results to the thinning of the sealant layer under sealing pressure at temperatures above sealant melting temperature. Mihindukulasuriya and Lim [10] used COMSOL multiphysics software to model heat transfer in heat sealing of two single-layer LLDPE films. During heat sealing, only the upper jaw was heated and a silicon rubber jaw was used for the bottom side as an insulator. They also applied Kapton® tape to the upper jaw to avoid sticking of the film to the jaw. Similar to Meka and Stehling [9], they did not consider variations of density, specific heat, and thermal conductivity with temperature.

As temperature of polymer film changes significantly during the heat sealing process [14], considering temperature dependency of the materials property is very important to achieve a reliable model. Kanani Aghkand et al. [3] used COMSOL multiphysics to model the heat sealing of film samples with PA/PE-*g*-MA/mLLDPE structure. They considered dependency of material properties by measuring temperature variations of density (ρ), specific heat (C_p), and thermal conductivity (k). Their results are shown in Figures 6.5–6.7.

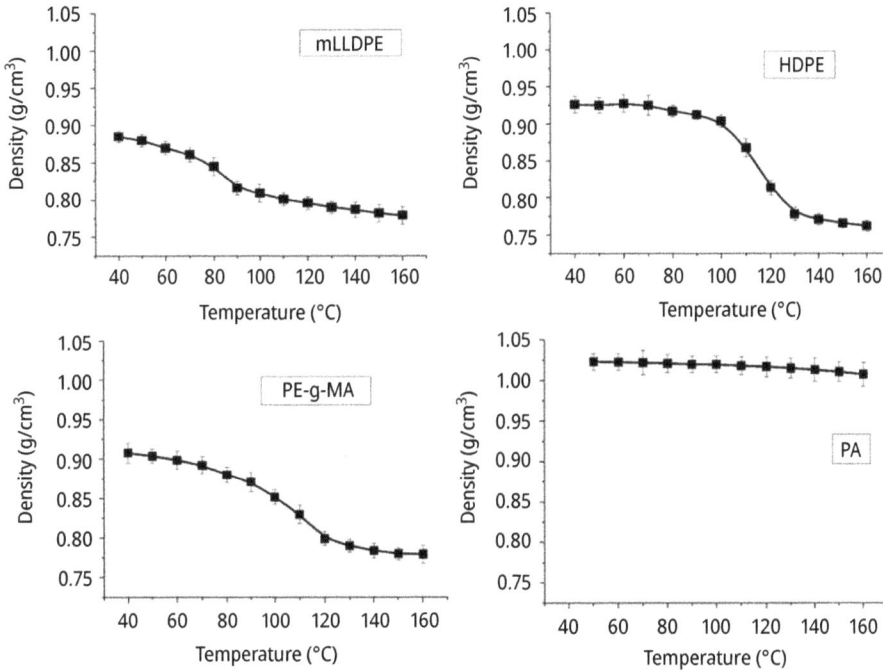

Figure 6.5: Variations of density of typical mLLDPE, HDPE, PE-*g*-MA, and PA.

As shown in Figure 6.5, density of polyethylene-based sealant materials and tie (PE-*g*-MA) decreased considerably after melting. This is attributed to the melting of ordered crystalline regions which increases free volume and reduces density. On the other hand, density variation of PA (the abuse layer) was within the range of experimental error and can be considered as constant.

Figure 6.6 shows that increasing temperature increased thermal conductivity of mLLDPE and PE-*g*-MA but reduced thermal conductivity of HDPE. Variations of thermal conductivity by increasing temperature can be explained by considering two opposing involved mechanisms [15–17]: (i) increasing the free volume by increasing temperature which reduces thermal conductivity; (ii) increasing segmental mobility which increases thermal conductivity. In highly crystalline HDPE, free volume increases significantly by

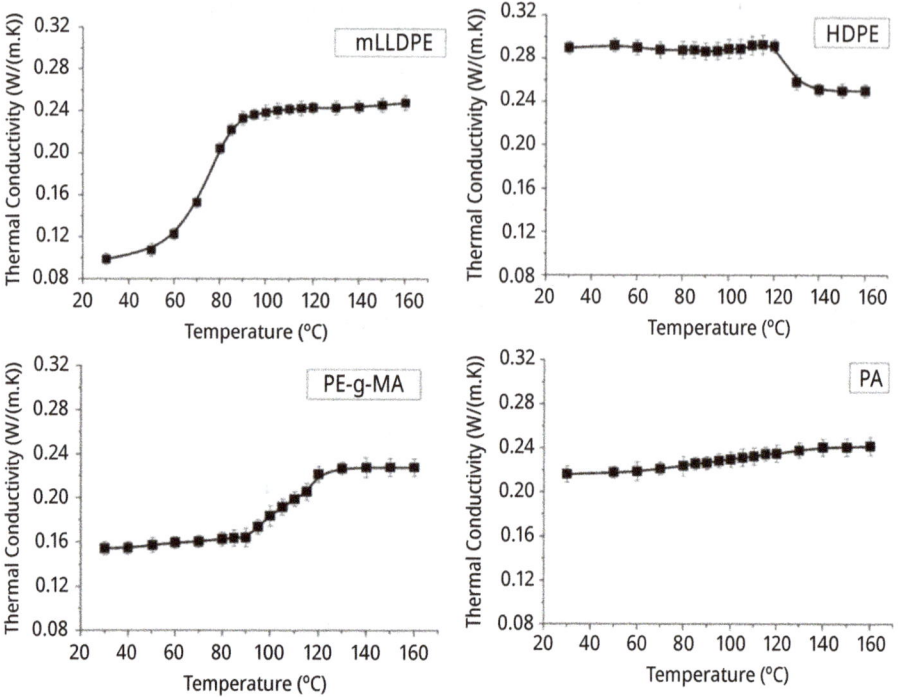

Figure 6.6: Variations of thermal conductivity of mLLDPE, HDPE, PE-*g*-MA, and PA by temperature.

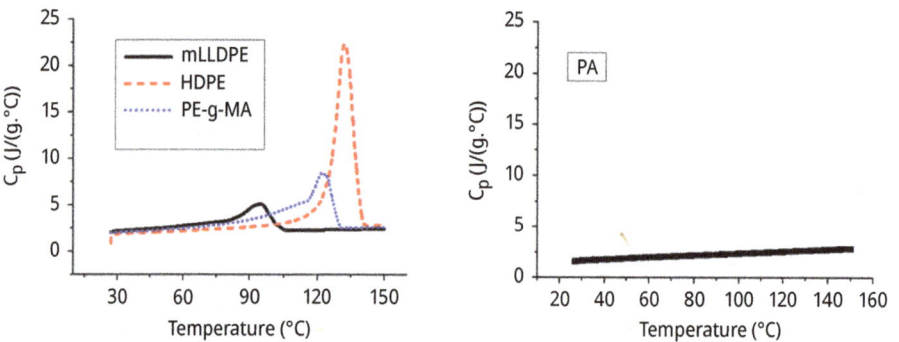

Figure 6.7: Temperature-modulated DSC (TMDSC) results showing the variations of specific heat of mLLDPE, HDPE, PE-*g*-MA, and PA.

melting crystals, and as a result, the first mechanism becomes dominant and reduces the thermal conductivity with temperature. On the other hand, in lower crystallinity mLLDPE and PE-*g*-MA, the change in the segmental mobility is dominant and, therefore, thermal conductivity of these materials increases by increasing temperature.

Specific heat measured by temperature-modulated DSC (TMDSC), Figure 6.7 shows a peak with a maximum at melting temperature. It should be noted that as DSC determines C_p based on the difference between the heat applied to the sample and the reference pans [18, 19], the change in C_p in the phase transition region indicates the heat absorbed by the material due to the phase transition. However, Kanani Aghkand et al. [3] showed that using the obtained values of C_p from TMDSC in the phase transition region allows considering the heat absorbed by the material during phase transition and, consequently, the difference between melting behavior of materials. This can be clearly seen in Figure 6.8 by comparing interface temperature of HDPE film at different jaw temperatures below and above HDPE melting temperature.

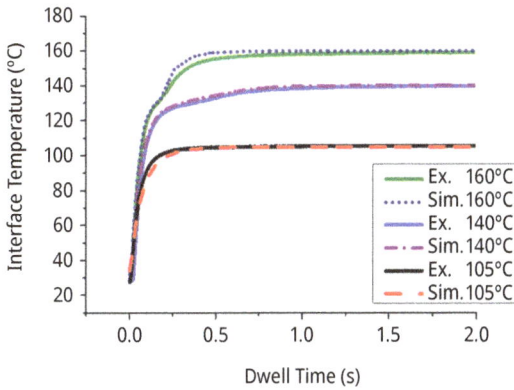

Figure 6.8: Comparison of experimental (Ex.) and simulation (Sim.) results for the interface temperature in heat sealing of HDPE at different jaw temperatures. Melting temperature of HDPE was 132 °C [3].

It can be seen that melting of HDPE causes a delay in heat transfer and results in longer times required for the interface temperature to reach jaw temperature. In the case of variation of C_p of PA with temperature, as the heat sealing is done at much lower temperature than its melting temperature, only a smooth linear increase in C_p of PA can be seen in the studied temperature range.

6.2 Modeling of squeeze-out flow

Channeling in the seal area is a common defect which can be due to nonoptimized sealing conditions such as temperature, pressure, or dwell time [20–22]. Channeling can occur in the wrinkled area, three-point junctions, or in the presence of contaminations in the seal area. Selecting sealants with good caulkability is a promising approach in eliminating channeling in wrinkles, junctions, or in the presence of contaminations [20, 22]. Caulkability is defined as the ability of a sealant to flow under the sealing pressure [22]. This phenomenon is also known as the squeeze-out flow or in short, SOF [21–23]. Caulkability is a necessary characteristic especially in the packaging of powders, grainy or shredded products such as coffee powder or shredded cheese that have a high chance of seal area contamination due to product residues [24, 25]. Despite the significant importance of SOF in heat sealing, few works have been done on the effects of sealant material properties and sealing condition on SOF in heat sealing process. Ward and Li [20] examined SOF in a wide range of different sealant materials but could not find a relation between SOF and zero-shear viscosity of sealant materials. Some previous works have also examined caulkability by attempting to correlate it to seal strength [24]; however, as the seal strength is affected by many material properties and sealing conditions, this approach cannot provide a reliable tool in examining the caulkability of sealants.

The high cost of industrial trials and the importance of SOF indicate the significant importance of modeling SOF during heat sealing. Squeeze flow of liquids between solid surfaces has been studied extensively due to its wide range of applications [21, 26–37]. Analytical models have been presented to estimate SOF in different geometries such as parallel rectangular plates [38], circular disks [34, 39], or even nonparallel surfaces [40] for both Newtonian and non-Newtonian fluids [41–43]. The previous works on modeling of SOF in the literature can be categorized into two main groups: (i) constant volume squeeze mode [31, 39] and (ii) constant contact area squeeze mode [34, 38]. In the first group, the liquid does not fill the gap between plates and the contact area between solid surfaces and the liquid increases during squeezing, but liquid is not squeezed out. This type of SOF is commonly seen in compression molding of polymer melts. In the constant contact area mode, the liquid always fills completely the gap between plates and squeezing of the plates leads to squeeze out of the liquid. SOF in heat sealing is an example of constant contact area mode. Most of the previous studies on squeeze flow in constant contact area category focused on SOF between two parallel circular disks in cylindrical coordination with very thick polymeric layer between them. Therefore, those studies cannot be directly applied to the squeeze flow within sealant layers in heat sealing. In order to determine the squeeze-out flow, authors commonly model two parallel plates filled with polymer and separated with an initial distance. The top plate approaches the bottom one under a pressure equal to the sealing pressure. The models predict the decrease in the gap between plates to determine the materials pulled out of the gap. Shuler and Advani [44] modeled transverse squeeze-out flow of a polymeric liquid between two rectangular plates with length of 2 L and width of W and found that for a Newtonian fluid with viscosity of η, under an applied force (F), the rate of gap thickness reduction ($\dot{h}$) can be estimated as

$$F = -8\,\eta W \frac{L^3}{h^3}\frac{dh}{dt} \qquad (6.7)$$

After integration, variation of the gap with time can be determined as

$$\frac{h}{h_0} = \left(\frac{Ph_0^2}{2L^2}t + 1\right)^{-1/2}$$ (6.8)

Equation (6.8) predicts that the squeeze-out flow depends on time with a power of −1/2. Morris and Scherer [45, 46] used the same equation for modeling squeeze-out flow but found considerable error in determining initial and final thickness (h and h_0) of the films. They proposed examining the change in the thickness at the edge of the seal bar to study squeeze-out flow. They reported that the squeezed-out region had a shape close to a trapezoid similar to the one schematically shown in Figure 6.9.

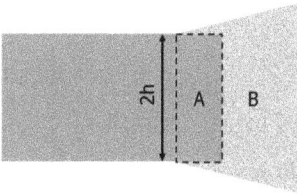

Figure 6.9: Schematic showing the method used by Morris and Scherer [45] for the estimation of squeeze-out flow in heat sealing of two films with thickness of h. B is the squeeze-out area (which includes A) and A shows a portion of the film cross-section overlapped with the squeeze-out area (or B).

They proposed that the amount of squeeze out can be determined using the following equation:

$$\% \text{ Squeeze out} = \frac{2(B - A)}{2h_0 W} \times 100$$ (6.9)

Finally, they compared experimental data and model predictions and found a poor agreement between model predictions and experimental results. The observed discrepancy between experimental and their model predictions was attributed to the simplifications assumptions such as constant temperature (no heat transfer was considered in their model).

Grewell and Benator [47] proposed that squeeze-out flow can be modeled by considering melting of surface asperities at the interface of two films. They assumed surface asperities as many small, identical cylinders on the surface of two parallel plates and considered a Newtonian fluid and constant material properties. Finally they obtained the following equation for the gap reduction by dwell time:

$$\frac{h}{h_0} = \left(\frac{16Fh^3}{3ar_0^4}t - 1\right)^{-1/4}$$ (6.10)

where h_0, h, F, r_0, and a are initial gap, gap opening at time t, the force applied by the upper plate, the radius of cylinders (surface asperities), and the thermal diffusivity, respectively. Equation (6.10) predicts that gap opening variation should be proportional to $t^{-\frac{1}{4}}$. However, this equation does not consider the effect of viscosity of the liquid. The authors did not present experimental results to examine their model predictions.

Levy et al. [48] examined the squeeze out of a molten thermoplastic between two rectangular plates using COMSOL multiphysics and compared their simulation results with experimental measurements. They used lubricating assumption and achieved the following set of analytical equations to predict the velocity fields:

$$\frac{d}{dy}\left(\eta\frac{du}{dy}\right) = -\frac{dP}{dx} \tag{6.11}$$

$$\int_{-h/2}^{h/2} u\,dy = x.V_{up} \tag{6.12}$$

$$\int_{-L/2}^{L/2} P\,dx = F/W \tag{6.13}$$

where u is the horizontal velocity that varies in the thickness (y) and horizontal (x) directions, P is the applied pressure to the top plate, V_{up} is the upper plate velocity, W is the depth of the sample in the third direction, and F is the applied force. Their simulation results are shown in Figure 6.10. It can be seen that the experimental values are always lower than the simulation and analytical model predictions. This can be attributed to the squeeze-out flow of the material in the third direction that was not considered in their 2D model. Their result show a dependency of $t^{-0.2}$, $t^{-0.4}$, and $t^{-0.7}$ for applied forces of 445, 1,334, and 2,222.4 N. Using the surface area of the sample (50 × 50 mm), the applied pressures can be calculated as 178, 536, and 889.6 kPa, respectively. These experimental results show that the time dependency of the thickness variation increases by increasing the applied force (pressure). The effect of pressure on time dependency of squeeze out has not been considered in the previous models in the literature.

Kanani Aghkand and Ajji [49] examined the effect of sealant material thickness, sealing condition, and sealant viscosity on SOF in film with a structure similar to the one shown in Figure 6.11.

They used a new image analysis approach to quantify SOF in which the film thickness was traced back into the sealed area until the traced lines reach each other in the center (Figure 6.12). These lines reach each other in the center at a point where sealed area should begin in the absence of SOF.

In the absence of SOF, the traced lines reach each other at the point where the sealed area begins (Figure 6.12(a1)). When the samples exhibit SOF (b1 and c1 in Figure 6.12), the traced lines reach each other inside of the sealed area. To calculate the amount of SOF, the area of the highlighted regions between traced lines in Figure 6.12 (b, c) was multiplied by the width of the sealed strips. As SEM measurements were done at room temperatures, the authors multiplied the calculated volume from image

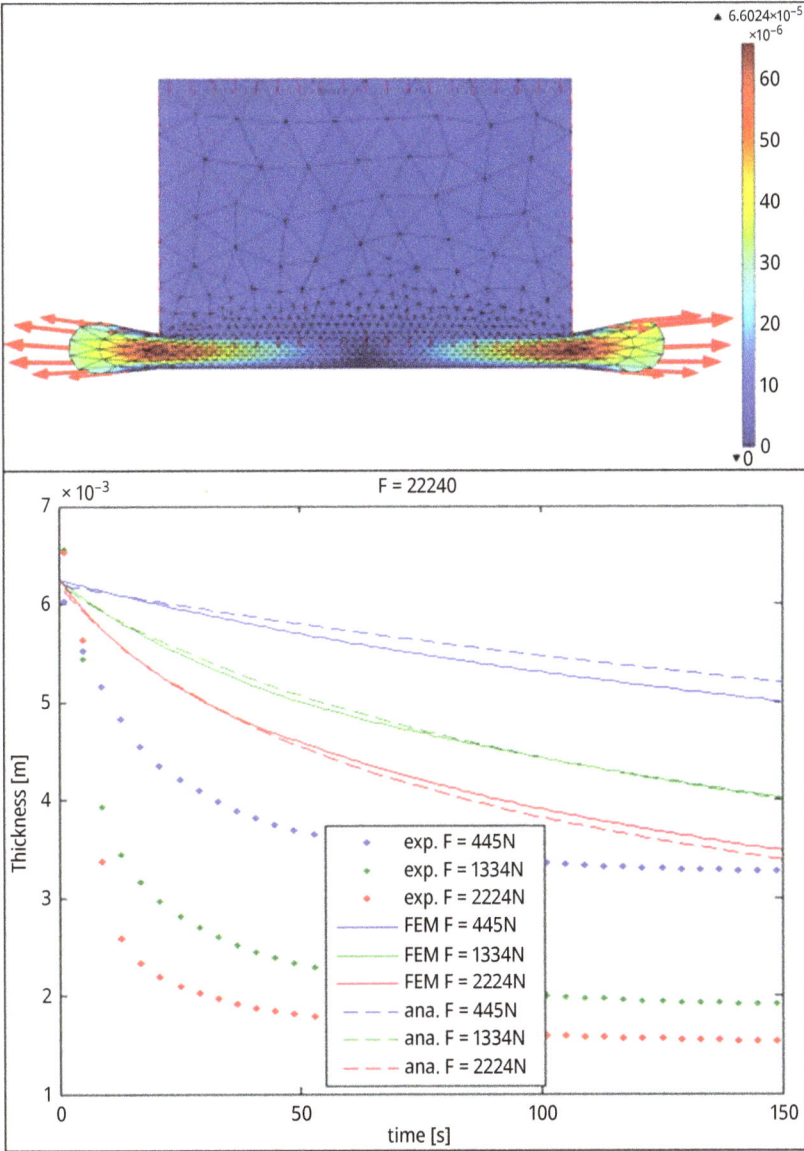

Figure 6.10: Simulation of squeeze-out flow of a thermoplastic material from a gap between two parallel plates. (Top) velocity field at t =150 s with applied force F =1,335 N and (Bottom) comparison of experimental results (shown as "exp") with Comsol simulation results (shown as "FEM") and prediction of numerical solution of analytical equations (shown as "ana"). F indicates the applied force in each case (reproduced with the permission of P. Hubert [48]).

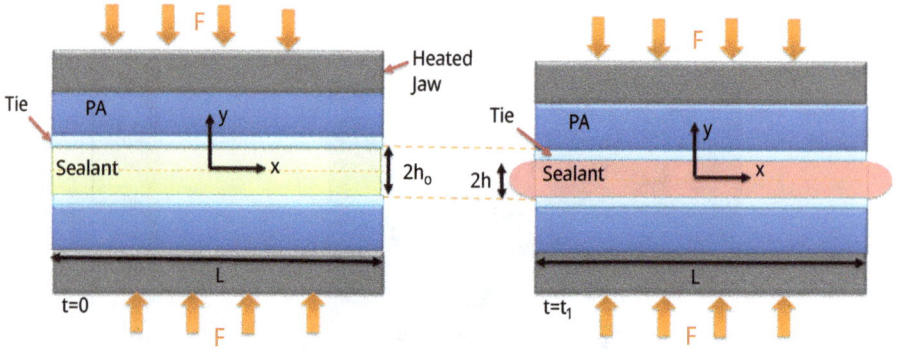

Figure 6.11: Squeeze flow during heat sealing process: (left) before and (right) after squeeze flow [49].

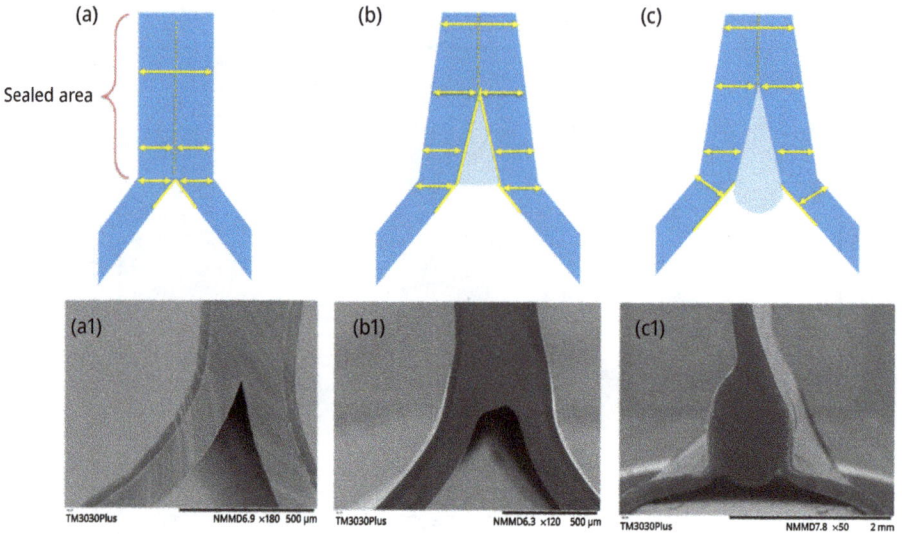

Figure 6.12: Schematic and SEM images of different possible shapes of the cross-section of sealed area in 130 μm sealants: (a) and (a1): no SOF ($P = 0.2$ MPa, $t = 1$ s, $T = 140$ °C); (b) and (b1) low SOF ($P = 0.2$ MPa, $t = 2$ s, $T = 140$ °C); (c) and (c1): high SOF or poly ball ($P = 3$ MPa, $t = 3$ s, $T = 140$ °C) [49].

analyses by the ratio of solid/melt densities. The authors presented one-dimensional models based on power-law fluid and Carreau–Yasuda fluid models based on the coordination system shown in Figure 6.11. Using a quasi-steady-state assumption, they could achieve the following analytical equation for the power-law model:

$$\frac{h_0}{h} = \left(1 + t\frac{4(n+1)}{L(2n+1)} h_0^{\frac{1}{n}+1} \left(\frac{2F}{mL^2W}\right)^{1/n}\right)^{\frac{n}{n+1}} \qquad (6.14)$$

where n is the power-law index and m is the consistency index. Using the Carreau–Yasuda fluid model, they obtained the following nonlinear differential equation:

$$\frac{\partial P}{\partial x} = \frac{\partial}{\partial y}\left(\eta_0\left(1+\left(-\lambda\frac{\partial V_x}{\partial y}\right)^a\right)^{\frac{n-1}{a}}\frac{\partial V_x}{\partial y}\right) \tag{6.15}$$

where P is the pressure inside the molten sealant. Due to the lack of an explicit analytical solution for equation (6.15), they used finite difference method (FDM) to solve this model. The authors compared the models prediction with experimental results. They showed that in a film with thick 130 µm sealant layer, increasing temperature from 105 to 140 °C increased significantly SOF (Figure 6.13).

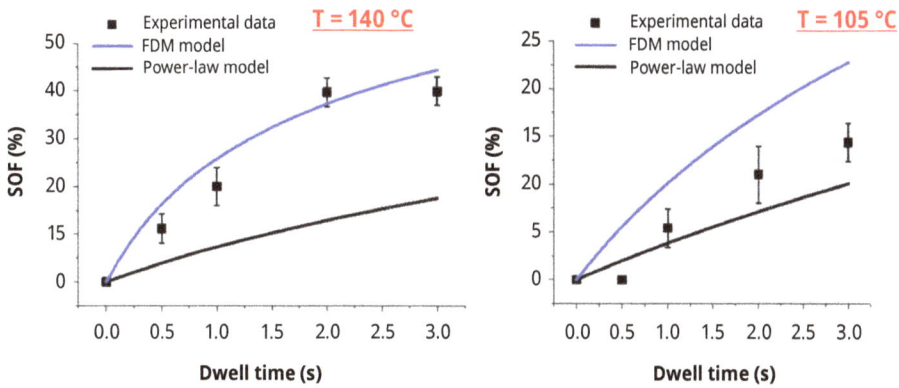

Figure 6.13: Comparing SOF at different dwell times with the predictions of the FDM and power-law models. All samples were sealed at sealing pressure of 3 MPa [49].

They found that the FDM model could predict much better the experimental data but still required more precision to be considered as a predictive model. The authors used heat transfer modeling and showed that heat transfer induces a lag in SOF. Figure 6.14 shows the heat transfer model predictions for temperature distribution within the sealant layer at different jaw temperatures and dwell times (Figure 6.14 (b, c)). Based on the predicted temperature profiles and by considering peak melting temperature at the criterion for melting, they could build a melt front versus heating time curve for each sealing temperature as shown in Figure 6.14(c). They then considered the time required for complete melting of the sealant layer as a horizontal shift factor for their FDM model and showed that using this approach, the FDM model could predict well the SOF experimental data. The authors also examined the effect of sealing layer thickness and showed that reducing sealant thickness from 130 to 50 µm suppressed SOF. They also examined the effect of sealant viscosity in a film with 50 µm sealant layer and found that reducing sealant viscosity could enhance SOF only at high sealing pressure and time. The authors compared the effect of sealing pressure on SOF in films with 130 and 50 μm sealants and, as shown in Figure 6.15, found that increasing sealing pressure by 10 times could only improve SOF in thick sealant and had almost no effect on SOF. As can be seen, their proposed FDM model could predict the effect of SOF in both thick and thin sealants.

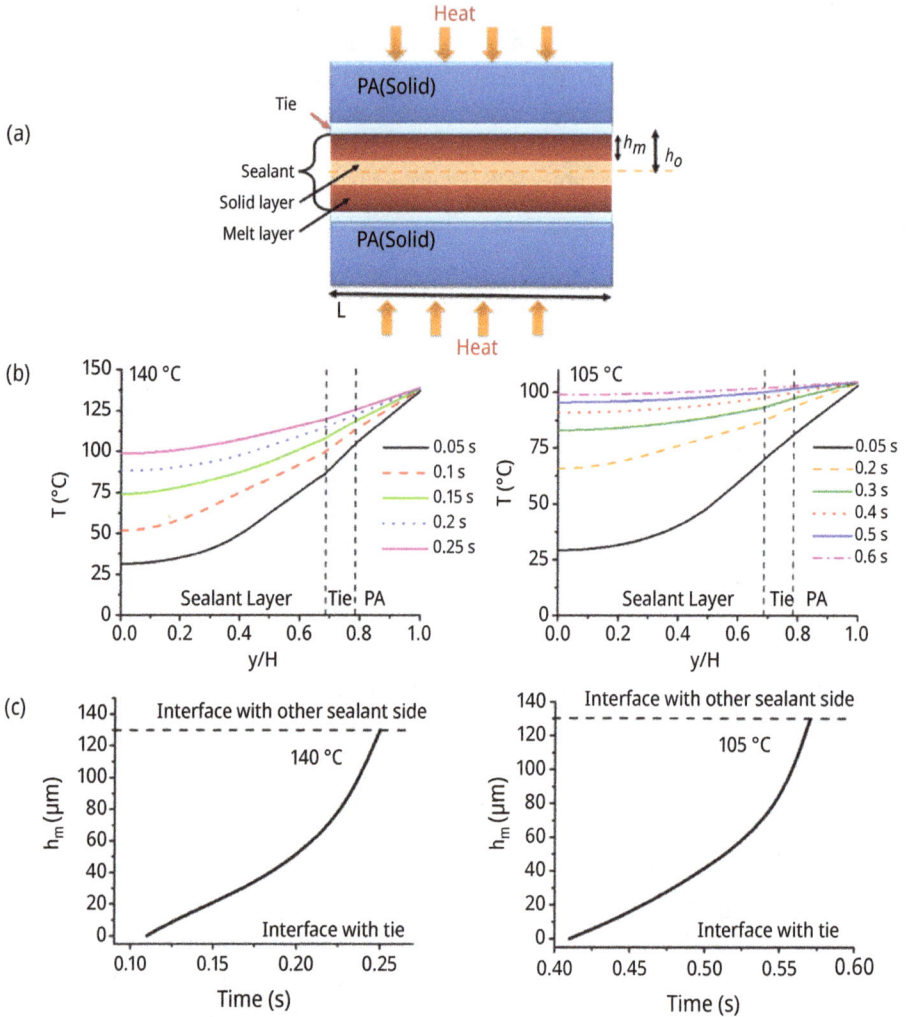

Figure 6.14: (a) Schematic of melt layer thickness evolution during heat sealing and (b) and (c) exact 3,131 melt layer thickness (h_m) at different times during heat sealing process at jaw temperatures of 140 and 105 °C. The total thickness of the sealant layer was 130 μm.

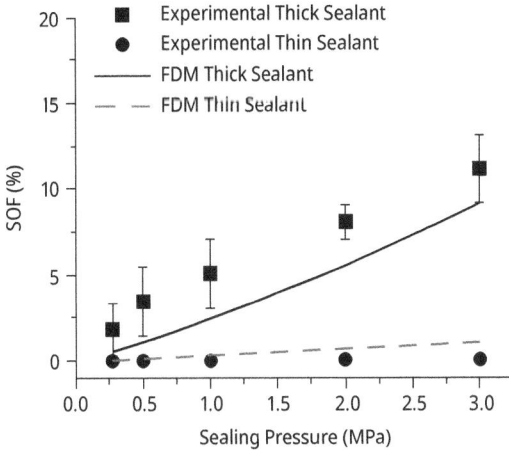

Figure 6.15: Effect of sealing pressure on SOF in films with thick (130 μm) and thin (50 μm) sealants. Heat sealing was done at 0.5 s and T=140 °C for all samples. The lines show shifted FDM model predictions [49].

In addition, using two-dimensional simulation of SOF in heat sealing using COMSOL multiphysics software and experimental data, they showed that the pressure profile inside the sealant is a nonlinear profile. Using the obtained sealant velocity profile, the authors estimated the shear rate during SOF and found that it falls within the transition region between the Newtonian and power-law regions. These results emphasizes on the importance of considering Carreau–Yasuda fluid model in prediction of SOF in heat sealing.

6.3 Modeling seal strength development

Predicting seal initiation temperature and final seal strength are two important goals of any heat sealing model. As mentioned in Chapter 2, both molecular diffusion and crystal formation after sealing contribute to the final seal strength while hot tack strength depends mainly on molecular diffusion. Crystallinity in the sealed area depends on molecular characteristics such as Mw, MWD, branching architecture, and branching density as well as cooling rate of the sealed area after sealing [50]. Therefore, prediction of seal strength requires much complex analysis compared with the hot tack strength. However, it should be considered that without enough molecular interdiffusion at the interface, even highly crystalline materials show poor seal strength. This indicates that the molecular diffusion is the main important factor, especially at the early stages of the seal strength development, while crystal formation enhances the strength by reducing mobility of the diffused chains. This indicates the significant importance of considering molecular interdiffusion in the prediction of seal initiation temperature. It should also be noted that, increasing initial crystallinity of the films can delay considerably the interface temperature by reducing heat transfer through heat absorption due to melting of these crystals. Considering all aforementioned, this section presents literature models for estimation of seal strength using molecular interdiffusion. These models focus mainly on the relation between fracture stress or fracture energy and time.

Wool [51] used the following rate law to establish a relation between the damage or void concentration per unit volume of the interface (C) and time:

$$-\frac{dC}{dt} = kt^n \tag{6.16}$$

where k is a temperature-dependent rate constant and n is the order of the healing process. Using this approach, the following empirical kinetic theory equation for crack healing was proposed to determine recovery or healing as a function of time:

$$\frac{\sigma(t)}{\sigma_\infty} \cong 1 - \frac{1-R_0}{(1+Kt)^{\frac{1}{1-n}}} \tag{6.17}$$

where σ_∞ is the bulk fracture stress and R_0 and K are constants. Equation (6.17) can predict the general sigmoidal behavior observed in seal strength development; however, order for damage disappearance (n) is unknown, and therefore, the order of healing process needs to be determined experimentally. This indicates that equation (6.17) is a strictly empirical equation that cannot predict the time dependency of the seal strength without experimental measurements.

Wool et al. [52] proposed the following scaling law, based on the reptation model, to relate dynamic properties of interfaces to their static (bulk) properties:

$$H(t) = H_\infty \left(\frac{t}{\tau_d}\right)^{r/4} \tag{6.18}$$

In this equation, $H(t)$ and H_∞ are any dynamic and static property of the interface, respectively. Moreover, r can be 1, 2, 3 . . . depending on the molecular property that is investigated. It has been shown that disentanglement time ($t < \tau_d$), fracture stress, and fracture energy of a polymer–polymer interface, in uniaxial tension, vary with time according to $t^{1/4}$ and $t^{1/2}$, respectively [52–56]. Therefore, it is likely to expect a dependency of $t^{1/2}$ for seal and tack strength in a heat sealing process.

Wool and O'Connor [57] developed another healing theory based on the molecular interdiffusion across the interface by considering two contributions for fracture stress (σ) as

$$\sigma = \sigma_0 + \sigma_d \tag{6.19}$$

where σ_0 is the stress due to wetting or surface attractions and σ_d is the stress due to chain interdiffusion across the interface. They assumed that the fracture stress originated from interdiffusion of polymer chains across the interface (σ_d) and is proportional to the number of the new constrains imposed on the diffused chain:

$$\sigma_d = qn = qn_0 X \tag{6.20}$$

where q, n, n_0, and X are a constant, the number of the new constrains, and the total number of constrains per unit volume of the virgin bulk material and the diffusion length of polymer chains, respectively. Using this assumption and by considering a diffusion initiation function $\dot{\psi}(t)$ and a wetting distribution function $\dot{\phi}(t)$, they obtained the following general healing function expressions for fracture stress and fracture energy:

$$\frac{\sigma(t)}{\sigma_\infty} = \left[\frac{\sigma_0}{\sigma_\infty} + \left(\frac{qn_0}{\sigma_\infty} (2Dt)^{1/4} \dot{\psi}(t) \right) \right] \dot{\phi}(t) \tag{6.21}$$

$$\frac{E(t)}{E_\infty} = \left[\frac{E_0}{E_{\sigma\infty}} + \left(\frac{q^2 n_0^2}{\sigma_\infty} (2Dt)^{1/2} \dot{\psi}(t) \right) \right] \dot{\phi}(t) \tag{6.22}$$

where E_∞ and D are the bulk fracture energy and the diffusion coefficient, respectively. The following conclusions can be made based on the proposed healing theory of Wool and O'Connor [57]:

i.   For instant wetting and instantaneous diffusion ($\dot{\phi}(t) = \dot{\psi}(t) = \delta(t)$, where $\delta(t)$ is the step function):

$$\sigma \sim t^{1/4} \text{ and } E \sim t^{1/2}$$

ii. For constant wetting rate and diffusion ($\dot{\phi}(t) = a.t$, $\dot{\phi}(t) = \delta(t)$):

$$\sigma \sim t + t^{5/4} \text{ and } E \sim t + t^{3/2}$$

iii. For constant wetting rate and small diffusion rate ($\dot{\phi}(t) = at$ and $D \to 0$):

$$\sigma, E \sim t$$

Considering the timescale in the heat sealing process, instant wetting and diffusion assumption are likely to be valid, and therefore, the seal strength is expected to obey $t^{1/2}$ dependency. It should be noted that the presented healing theory of Wool and O'Connor can also be used to predict the effect of molecular weight on seal strength by considering $D \sim 1/M_W$.

Moffitt [58] proposed an integral form equation for evolution of peel strength with time. By comparing model predictions with literature data, he reported that considering time dependency of wetting process is necessary in the prediction of seal strength.

Morris [59] proposed to use the Brownian motion concept and the terminal relaxation time to estimate the diffusion length using the following equation [59]:

$$W^* = \frac{W}{R_g} = \frac{4\sqrt{Dt}}{4\sqrt{Dt_{d,avg}}} = \sqrt{\frac{t}{t_{d,avg}}} \tag{6.23}$$

where W^*, W, R_g, D, and $t_{d,avg}$ are dimensionless penetration distance, penetration distance at time t, the radius of gyration of the polymer chain, diffusion coefficient, and the average terminal relaxation time. He used a heat transfer model to estimate the interface temperature at each time and determined the average terminal relaxation time using the following equation:

$$\frac{1}{t_{d,avg}} = \frac{1}{n}\sum_{i=1}^{i=n}\frac{1}{t_{d,i}} \tag{6.24}$$

A critical penetration distance called W_c^* was defined where the penetration length corresponds with the onset of the seal strength plateau region. By assuming direct correlation between dimensionless diffusion length and seal strength, Morris fitted the variation of plateau initiation temperature with dwell time in two different cases to determine W_c^* as the fitting parameter. He found that in both cases, W_c^* was lower than 0.1. This indicates that the seal formation was done at much smaller length scale than R_g of the polymer chain. By comparing experimental results and model predictions, they proposed that a direct relation between W_c^* and the seal strength could be established.

References

[1]   Ward, D., *Heat Transfer Modelling in Multilayer Films Used for Flexible Packaging: Hot Tack and Thermoforming Application Case Studies*. In *TAPPI*. 2018.

[2]   Morris, B.A., *Heat Seal, in the Science and Technology of Flexible Packaging: Multilayer Films from Resin and Process to End Use*. 2017, Elsevier Inc.

[3]   Kanani Aghkand, Z., et al., *Simulation of heat transfer in heat-sealing of multilayer polymeric films: Effect of process parameters and material properties*. Industrial & Engineering Chemistry Research, 2018.

[4]   Cooper, M., B. Mikic, and M. Yovanovich, *Thermal contact conductance*. International Journal of Heat and Mass Transfer, 1969. **12**(3): p. 279–300.

[5]   Mikić, B., *Thermal contact conductance; theoretical considerations*. International Journal of Heat and Mass Transfer, 1974. **17**(2): p. 205–214.

[6]   Song, S., M. Yovanovich, and F. Goodman, *Thermal gap conductance of conforming surfaces in contact*. Journal of Heat Transfer, 1993. **115**(3): p. 533–540.

[7]   Sridhar, M. and M. Yovanovich, *Review of elastic and plastic contact conductance models-Comparison with experiment*. Journal of Thermophysics and Heat Transfer, 1994. **8**(4): p. 633–640.

[8]   Wahid, S.M. and C. Madhusudana, *Gap conductance in contact heat transfer*. International Journal of Heat and Mass Transfer, 2000. **43**(24): p. 4483–4487.

[9]   Meka, P. and C. Stehling Ferdinand, *Heat sealing of semicrystalline polymer films. I. Calculation and measurement of interfacial temperatures: Effect of process variables on seal properties*. Journal of Applied Polymer Science, 1994. **51**(1): p. 89–103.

[10]   Mihindukulasuriya, S.D. and L. T. Lim, *Heat sealing of LLDPE films: Heat transfer modeling with liquid presence at film–film interface*. Journal of Food Engineering, 2013. **116**(2): p. 532–540.

[11]   Song, S., M. Yovanovich, and K. Nho, *Thermal gap conductance-Effects of gas pressure and mechanical load*. Journal of Thermophysics and Heat Transfer, 1992. **6**(1): p. 62–68.

[12]   Zheng, J., et al., *Measurements of interfacial thermal contact conductance between pressed alloys at low temperatures*. Cryogenics, 2016. **80**: p. 33–43.

[13]   Kanani Aghkand, Z., et al., *Simulation of Heat Sealing Process, in Polymer Processing Society (PPS) Americas*. 2018, Boston.

[14]   Hishinuma, K., *Heat Sealing Technology and Engineering for Packaging: Principles and Applications*. 2009, DEStech Publications, Inc.

[15]   Choy, C.L., *Thermal conductivity of polymers*. Polymer, 1977. **18**.

[16]   Dos Santos, W.N., J.A. De Sousa, and R. Gregorio, *Thermal conductivity behaviour of polymers around glass transition and crystalline melting temperatures*. Polymer Testing, 2013. **32**(5): p. 987–994.

[17]   Dashora, P. and G. Gupta, *On the temperature dependence of the thermal conductivity of linear amorphous polymers*. Polymer, 1996. **37**(2): p. 231–234.

[18]   Höhne, G., W.F. Hemminger, and H.-J. Flammersheim, *Differential Scanning Calorimetry*. 2013, Springer Science & Business Media.

[19]   Gill, P., S. Sauerbrunn, and M. Reading, *Modulated differential scanning calorimetry*. Journal of Thermal Analysis, 1993. **40**(3): p. 931–939.

[20]   Ward, D. and M. Li, *Seal-through-contamination and "caulkability"an evaluation of sealants' ability to encapsulate contaminants in the seal area*. In *TAPPI PLACE Conference*. 2016, Fort Worth, Texas.

[21]   Morris, B.A., *The Science and Technology of Flexible Packaging: Multilayer Films from Resin and Process to End Use*. 2016, Elsevier Science & Technology Books.

[22]   Morris, B.A. and J.M. Scherer, *Modeling and experimental analysis of squeeze flow of sealant during hot bar sealing and methods of preventing squeeze-out*. Journal of Plastic Film and Sheeting, 2015. **32**(1): p. 34–55.

[23]   Morris, B.A., *Predicting the performance of ionomer films in heat-seal processes*. Paper, Film and Foil Converter, 2003. **77**(3): p. 207.

[24]   Mesnil, P., et al., *Seal through contamination performance of metallocene plastomers*. In *TAPPI Polymers, Laminations & Coatings Conference*. 2000, Chicago.

[25]   Bamps, B., et al., *Evaluation and optimization of seal behaviour through solid contamination of heat-sealed films*. Packaging Technology and Science, 2019. **32**(7): p. 335–344.

[26]   Picher-Martel, G.-P., A. Levy, and P. Hubert, *Compression moulding of carbon/PEEK randomly-oriented strands composites: A 2D finite element model to predict the squeeze flow behaviour*. Composites Part A: Applied Science and Manufacturing, 2016. **81**: p. 69–77.

[27]   Levy, A., G.P. Martel, and P. Hubert, *Modeling the Squeeze Flow of a Thermoplastic Composite Tape during Forming*, in *COMSOL Conference*. 2012, Boston, USA.

[28]   Hou, J., et al., *An analysis of the squeeze-film lubrication mechanism for articular cartilage*. Journal of Biomechanics, 1992. **25**(3): p. 247–259.

[29]   Mueller, C., et al., *Heat sealing of LLDPE: Relationships to melting and interdiffusion*. Journal of Applied Polymer Science, 1998. **70**(10): p. 2021–2030.

[30]   Ting, G., *Polymer Welding: Strength Through Entanglements*. 2012.

[31]   Grewell, D. and A. Benatar, *Welding of plastics: Fundamentals and new developments*. International Polymer Processing, 2007. **22**(1): p. 43–60.

[32]   Grewell, D. and A. Benatar, *Coupled temperature, diffusion and squeeze flow model for interface healing predictions*. In *ANTEC*. 2006. p. 2205–2210.

[33]   Engmann, J., C. Servais, and A.S. Burbidge, *Squeeze flow theory and applications to rheometry: A review.* Journal of Non-Newtonian Fluid Mechanics, 2005. **132**(1): p. 1–27.

[34]   Leider, P.J. and R.B. Bird, *Squeezing flow between parallel disks. I. Theoretical analysis.* Industrial & Engineering Chemistry Fundamentals, 1974. **13**(4): p. 336–341.

[35]   Chan, T.W. and D.G. Baird, *An evaluation of a squeeze flow rheometer for the rheological characterization of a filled polymer with a yield stress.* Rheologica Acta, 2002. **41**(3): p. 245–256.

[36]   Suwonsichon, T. and M. Peleg, *Rheological characterisation of almost intact and stirred yogurt by imperfect squeezing flow viscometry.* Journal of the Science of Food and Agriculture, 1999. **79**(6): p. 911–921.

[37]   Campanella, O. and M. Peleg, *Squeezing flow viscosimetry of peanut butter.* Journal of Food Science, 1987. **52**(1): p. 180–184.

[38]   Gupta, P.S. and A.S. Gupta, *Squeezing flow between parallel plates.* Wear, 1977. **45**(2): p. 177–185.

[39]   Dienes, G. and H. Klemm, *Theory and application of the parallel plate plastometer.* Journal of Applied Physics, 1946. **17**(6): p. 458–471.

[40]   Sherwood, J.D., *Squeeze flow of a power-law fluid between non-parallel plates.* Journal of Non-Newtonian Fluid Mechanics, 2011. **166**(5): p. 289–296.

[41]   Wilson, S.D.R., *Squeezing flow of a Bingham material.* Journal of Non-Newtonian Fluid Mechanics, 1993. **47**: p. 211–219.

[42]   Sherwood, J.D. and D. Durban, *Squeeze flow of a power-law viscoplastic solid.* Journal of Non-Newtonian Fluid Mechanics, 1996. **62**(1): p. 35–54.

[43]   Lian, G., et al., *On the squeeze flow of a power-law fluid between rigid spheres.* Journal of Non-Newtonian Fluid Mechanics, 2001. **100**(1): p. 151–164.

[44]   Shuler, S.F. and S.G. Advani, *Transverse squeeze flow of concentrated aligned fibers in viscous fluids.* Journal of Non-Newtonian Fluid Mechanics, 1996. **65**(1): p. 47–74.

[45]   Morris, B.A. and J.M. Scherer, *Modeling and experimental analysis of squeeze flow of sealant during hot bar sealing and methods of preventing squeeze-out.* Journal of Plastic Film & Sheeting, 2015. **32**(1): p. 34–55.

[46]   Morris, B.A. and J.M. Sherer, *Preventing squeeze-out of sealant during hot bar sealing: Modeling and experimental insights*, in *72nd Annual technical conference of the Society of Plastics Engineers (ANTEC 2014).* 2014, Las Vegas, TX, US.

[47]   Grewell, D. and A. Benatar. *Coupled temperature, diffusion and squeeze flow model for interfacial healing predictions.* In *Proceedings of the Annual Conference of the Society of Plastics Engineers (ANTEC).* 2006.

[48]   Levy, A., G.P. Martel, and P. Hubert. *Modeling the squeeze flow of a thermoplastic composite tape during forming.* In *Proceedings of the 8th annual COMSOL conference.* 2012.

[49]   Kanani Aghkand, Z. and A. Ajji, *Squeeze flow in multilayer polymeric films: Effect of material characteristics and process conditions.* Journal of Applied Polymer Science, 2021. Accepted.

[50]   Di Lorenzo, M.L. and C. Silvestre, *Non-isothermal crystallization of polymers.* Progress in Polymer Science, 1999. **24**(6): p. 917–950.

[51]   Wool, R.P., *Crack* healing in semicrystalline polymers, block copolymers and filled elastomers. In *Adhesion and Adsorption of Polymers*, L.-H. Lee, Editor. 2012, Springer Science & Business Media.

[52]   Wool, R., B.L. Yuan, and O. McGarel, *Welding of polymer interfaces.* Polymer Engineering & Science, 1989. **29**(19): p. 1340–1367.

[53]   Kim, Y.H. and R.P. Wool, *A theory of healing at a polymer-polymer interface.* Macromolecules, 1983. **16**(7): p. 1115–1120.

[54]   Wool, R.P. and K.M. O'Connor, *A theory crack healing in polymers.* Journal of Applied Physics, 1981. **52**(10): p. 5953–5963.

[55]   Whitlow, S.J. and R.P. Wool, *Diffusion of polymers at interfaces: A secondary ion mass spectroscopy study.* Macromolecules, 1991. **24**(22): p. 5926–5938.

[56] Bastien, L.J. and J.W. Gillespie, *A non-isothermal healing model for strength and toughness of fusion bonded joints of amorphous thermoplastics*. Polymer Engineering & Science, 1991. **31**(24): p. 1720–1730.

[57] Wool, R. and K. O'connor, *A theory crack healing in polymers*. Journal of Applied Physics, 1981. **52**(10): p. 5953–5963.

[58] Moffitt, R.D., *A mathematical model for the heat sealing of linear, semi-crystalline polymers*. In *ANTEC*. 2006.

[59] Morris, B.A., *Predicting the heat seal performance of ionomer films*. Journal of Plastic Film & Sheeting, 2002. **18**(3): p. 157–167.

Chapter 7
Multicomponent sealant films

Increasing demand for seal layers with wide range of properties and functionality and the limited capabilities of a single polymer component sealant have increased the application of multicomponent sealants in packaging industry. These multicomponent sealants can play various roles in a package including, but not limited to, modifying sealing, mechanical and rheological properties of sealant. Multicomponent sealant films are mainly produced from blending of multiple polymer components. Blending has been shown to be a promising method in achieving high-performance polymer materials. From sealing point of view, the main goals for using polymer blends as the multicomponent sealant layer can be summarized as (i) reducing sealing temperature, (ii) broadening the sealing window, (iii) possibility of asymmetric sealing, (iv) enhancing caulkability, (v) changing sealing behavior, and (vi) optimizing sealant cost. Before discussing sealants based on polymer blends, first a short review of thermodynamic and morphology of polymer blends is presented in the next section.

7.1 Thermodynamics of polymer blends

From the thermodynamic point of view, polymer blends can be categorized into three groups: miscible, partially miscible, and immiscible blends. Miscible polymer blends form one-phase mixture that is homogeneous in the molecular scale. Partially miscible blends form two phases in which each phase contains both polymer component molecules. Each of these phases is homogeneous at the molecular scale but contains different compositions of polymer components compared to the other one. The two phases have their boundaries known as the interface and can have different shape and size or in another word can have different morphology. Immiscible polymer blends form two phases in which each phase consists of only one polymer component. Like partially miscible blends, they also have interface between the phases and can have different morphologies. Figure 7.1 shows schematically the differences between these types of blends.

The thermodynamic of a polymer blend system can be described using the Gibbs's free energy of mixing (ΔG_m):

$$\Delta G_m = \Delta H_m - T\Delta S_m \qquad (7.1)$$

In this equation, ΔH_m and ΔS_m are the enthalpy and entropy of mixing, respectively. A miscible blend is formed when negative free energy of mixing is achieved ($\Delta G_m < 0$). Different equations for estimation of the enthalpy and the entropy of mixing have been discussed widely in literature (see reference [1] for example) and will not be discussed here. The entropy of mixing has been shown to change inversely with the

https://doi.org/10.1515/9781501524592-007

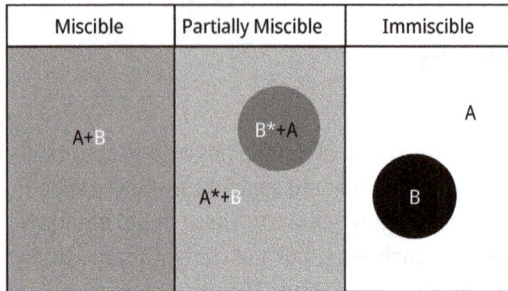

Figure 7.1: Different polymer blend types based on their miscibility.

molecular weight of components in binary mixtures [2–4]. Therefore, due to the very high molecular weights of polymers, the entropy of mixing (ΔS_m) in polymeric mixtures is typically negligible. In the absence of specific interactions such as hydrogen bonding [5], it has been shown that polymer mixtures exhibit a positive enthalpy of mixing ($\Delta H_m > 0$) [1]. Therefore, in most cases, mixing of polymer components results in a positive ΔG_m and the formation of an immiscible polymer blend. Miscible blends of different polyethylene are the most common miscible polymer blend examples used as sealant material in plastic film. For example, blending of low temperature sealing metallocene PE (mPE) with conventional PE is a common approach to reduce S.I.T. and improve seal strength while maintaining reasonable material cost. Najarzadeh et al. [47] did a systematic study on the effects of the addition of mPE with linear or long-chain branching (LCB) molecular structures on seal strength of conventional LLDPE or LDPE and showed that the blending approach is much effective when the matrix PE has LCB (Figure 7.2). The % shown in the figure indicates the increase in seal (or adhesion) strength based on the initial PE type, LLDPE (LL) or LDPE (LD).

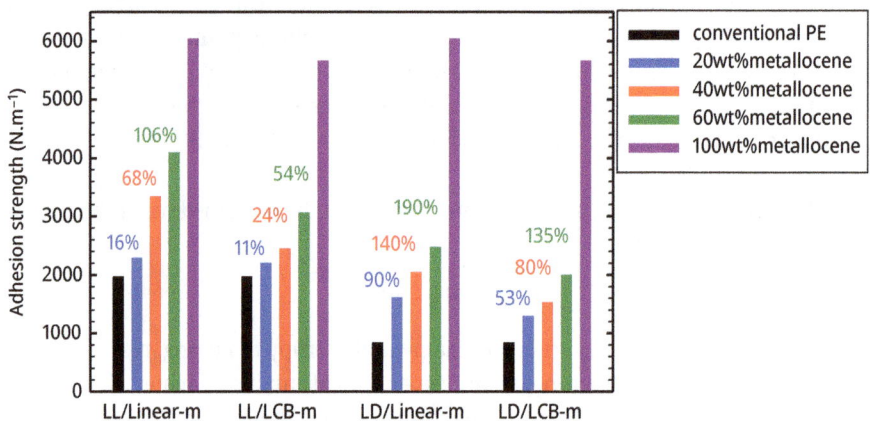

Figure 7.2: Effect of the addition of linear or LCB mPE on seal strength of LLDPE or LDPE [47].

They also studied the effect of addition of mPE with linear or LCB structures on hot tack of the same blends. Interestingly, they found that the addition of both mPE types broadened the hot tack window but only the addition of linear mPE could reduce hot tack initiation temperature.

As in most cases blending of polymers leads to the formation of immiscible polymer blends, our main focus in this chapter will be dedicated to immiscible polymer blends. In immiscible polymer blends, the interactions between the two phases occur at their boundaries or their interface. These interactions indicate the level of compatibility of the phases in the system which directly influences final properties. The level of interactions between two nonreactive phases can be expressed using the work of adhesion (W_{adh}) between them. The work of adhesion is the energy required for separating a unit area of the interface between two phases [6]. Wu [7] showed that W_{adh} can be estimated using the following two approaches:

(i) Harmonic mean

$$W_{adh} = 4 \left(\frac{\gamma_1^d \gamma_2^d}{\gamma_1^d + \gamma_2^d} + \frac{\gamma_1^p \gamma_2^p}{\gamma_1^p + \gamma_2^p} \right) \qquad (7.2)$$

(ii) Geometric mean

$$W_{adh} = 2 \left(\sqrt{\gamma_1^d \gamma_2^d} + \sqrt{\gamma_1^p \gamma_2^p} \right) \qquad (7.3)$$

where γ^d and γ^p are the contributions of dispersive and polar components in the surface tension of components 1 and 2, respectively.

The harmonic mean approach should be used for the estimation of the work of adhesion between two phases with similar polarities such as between polymeric materials. On the other hand, the geometric mean equation should be used for the estimation of the work of adhesion between two phases with significantly different polarities such as mixtures of polymers and inorganic solid materials.

Another important parameter in immiscible polymer blends is the interfacial tension between phases (γ_{12}) which is defined as the work required to increase the interfacial area between phases by a unit of area [6–8]. The interfacial tension between two polymer melts can be estimated using their surface tensions (γ_1 and γ_2) and the work of adhesion between them (W_{adh}) using the following equation [6, 7]:

$$\gamma_{12} = \gamma_1 + \gamma_2 - W_{adh} \qquad (7.4)$$

A high level of compatibility between phases results in a greater work of adhesion and, consequently, a lower interfacial tension. Different techniques have been used to measure experimentally the interfacial tension between polymers such as the breaking thread method [9], the fiber retraction method [10, 11], the pendant droplet method [12],

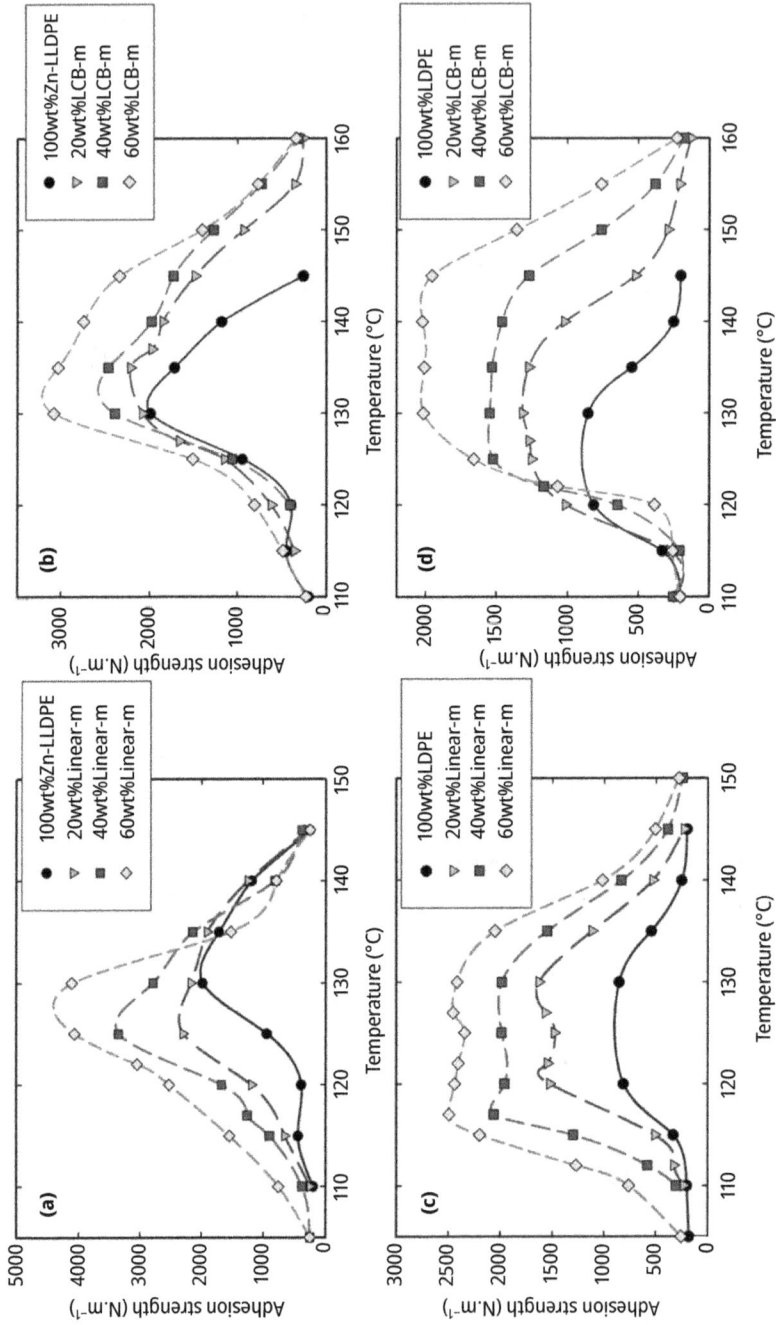

Figure 7.3: Effect of the addition of mPE with linear or long-chain branching (LCB) structures on hot tack of different polyethylene types [47].

and the rheological approach [13, 14]. The details of these techniques will not be reviewed here and can be found in the cited references.

7.2 Morphology of polymer blends

The morphology of immiscible polymer blends directly controls their final properties [15, 16]. It should be noted that the term morphology refers to the shape and size of the phases in the blend microstructure [17]. Different common morphologies observed in binary immiscible polymer blends are listed below in Figure 7.4 with their potential applications.

	Matrix-Dispersed			Co-Continuous
	Spherical	Fibrillar	Lamellar	
Morphology				
Application	Toughness	Strength	Barrier Properties	Conductivity

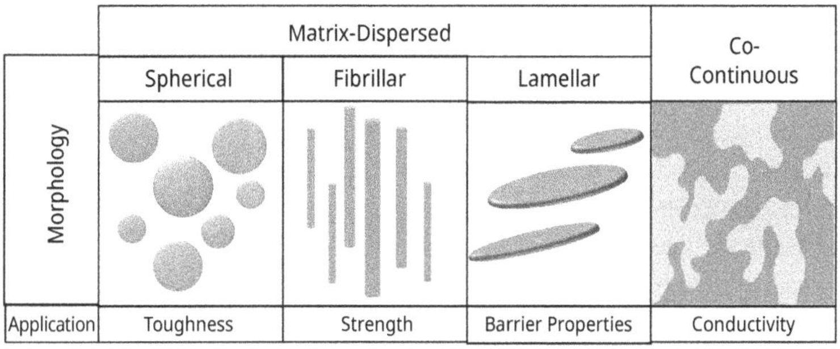

Figure 7.4: Common morphologies in binary immiscible polymer blends with their potential applications.

The final morphology of a polymer blend is the result of a competition between breakup and coalescence of dispersed phase during melt mixing [17, 18]. Breakup and coalescence phenomena are controlled by intrinsic properties of polymer phases such as their interfacial tension [19] and viscosity [20] as well as processing conditions such as shear rate and the flow field [21, 22]. Breakup of a droplet with radius R in a viscous matrix having a viscosity of η can be evaluated by considering the capillary number (Ca) that is defined as [23]

$$Ca = \frac{\dot{\gamma}}{\frac{\gamma_{12}}{R}} \tag{7.5}$$

The capillary number is the ratio of the viscous forces ($(\dot{\gamma})$) to the interfacial forces (γ_{12}/R). It is known that viscous forces tend to deform and eventually break up the dispersed droplets while the interfacial forces withstand droplet deformation. Droplet deformation and eventually breakup occurs above a certain capillary number known as the critical capillary number (Ca$_c$) where the viscous forces overcome the interfacial forces and can deform and eventually breakup the dispersed phase droplets. Figure 7.5

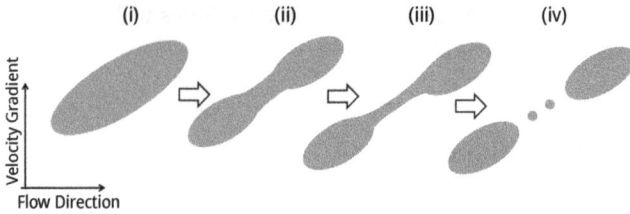

Figure 7.5: The deformation and breakup process of a dispersed phase droplet in a shear flow field: (i) to (iii) droplet deformation and (iv) droplet breakup and formation of satellite small droplets.

schematically shows deformation and breakup of a dispersed phase droplet in a viscous matrix under shear flow.

Coalescence of dispersed phase is another important phenomenon that competes with the droplet breakup. During coalescence, two dispersed phase domains approach each other and drain a film of the matrix trapped between them under shear forces applied by the surrounding medium. When the thickness of the film trapped between dispersed domains reaches a critical value, the film ruptures and two droplets merge into a larger droplet. According to the Laplace law, the pressure inside each droplet is inversely related to its radius; therefore, domains with smaller sizes have a higher internal pressure and are drained into the larger domain during coalescence [24]. The coalescence process is schematically shown below in Figure 7.6.

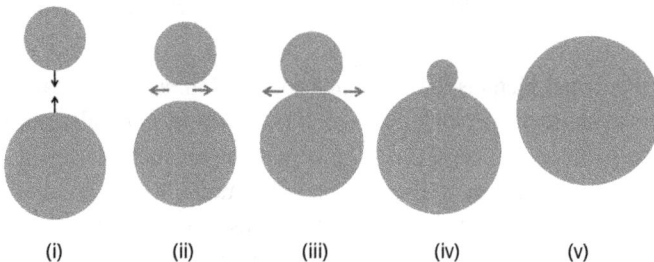

Figure 7.6: Schematic of coalescence process: (i) approaching droplets, (ii) deformation of contact area between droplets and formation of matrix film between them, (iii) draining of the trapped matrix film, (iv) matrix film rupture and merging of droplets, and (v) final merged droplets.

Different models have been proposed to explain the effect of different parameters on coalescence in a polymer blend. The probability of coalescence can be explained by considering the following equation [25, 26]:

$$P \approx \exp\left[-\frac{t_{\text{drain}}}{t_{\text{int}}}\right] \tag{7.6}$$

where t_{drain} and t_{int} are the draining time of the matrix film between two droplets and the time that two droplets are in contact before being moved away by the flow field. For example, it has been shown that increasing shear rate at low shear rates enhances coalescence while higher shear rates reduce the coalescence between dispersed phases [27]. It should be noted that reducing shear rate reduces both t_{drain} and t_{int} [28]. These results indicate that, at least in the studied system, the effect of shear rate on t_{int} was much significant than on t_{drain}.

In order to better understand the importance of the above-mentioned discussions, two examples of the importance of the interfacial tension and flow field are discussed here. Li et al. [19] studied the effect of interfacial tension on the morphology development in immiscible polymer blends. To this aim, they compared the morphology of a high interfacial tension blend of high-density polyethyelene (HDPE)/and polystyrene and a low interfacial tension blend of HDPE/styrene–ethylene–butylene–styrene. They found that the morphology of the dispersed phase in the high interfacial tension system is spherical droplets while in the low interfacial tension the dispersed phase exists in the form of elongated fiber-like morphology. This indicates easier droplet breakup in higher interfacial tension systems where interfacial forces are weak. In another study, Jalali Dil et al. [29] studied the effect of flow field on the morphology of low interfacial tension poly(lactic acid) and poly(butylene adipate-*co*-terephthalate) (PLA/PBAT) blends. They compared the morphology of PLA/PBAT films prepared at different blow-up ratios (BUR) in film blowing process and found that changing the flow filed from 1D (BUR ~ 1) to 2D (BUR = 3) changed the morphology of the dispersed PBAT phase from a fibrillar to a lamellar morphology. It was found that changing the morphology from fibrillar to lamellar reduced the seal initiation temperature of the film samples by increasing the surface area of low sealing temperature phase (PBAT) at the film surface.

These two examples clearly indicate the significant importance of controlling morphology of polymer blend sealants to achieve the final desired properties including sealing performance.

7.3 Surface morphology of immiscible polymer blend films

In addition to bulk properties, controlling the morphology and microstructure of polymer blends is critical in controlling their surface properties [15, 30]. As sealing properties are directly controlled by the surface morphology, controlling the surface morphology of a polymer blend is necessary in achieving the desired seal properties. From the thermodynamic view point, the surface morphology of sealant films right after exiting the extrusion die depends on the interactions of the blend components with the metallic inner surface of the extrusion die. The thermodynamics of the localization of the phases with respect to the surface of the die can be explained using the Harkins' equations [31, 32]:

$$\lambda_{\mathrm{SAB}} = \gamma_{\mathrm{SB}} - (\gamma_{\mathrm{SA}} + \gamma_{\mathrm{AB}}) \tag{7.7}$$

$$\lambda_{SBA} = \gamma_{SA} - (\gamma_{SB} + \gamma_{AB})$$ (7.8)

where λ, γ_{SB}, γ_{SA}, and γ_{AB} are the spreading coefficients, the interfacial tension between solid surface and phase B, the interfacial tension between solid surface and phase A, and the interfacial tension between two polymer phases, respectively. Figure 7.7 shows the different possible morphology scenarios depending on spreading coefficient values in a system with dispersed-matrix morphology. It is worth mentioning that when surface wetting by the minor component is preferred, a minimum amount of minor phase is needed to form a complete layer. Below such content, discrete domains of the minor phase are formed at the surface.

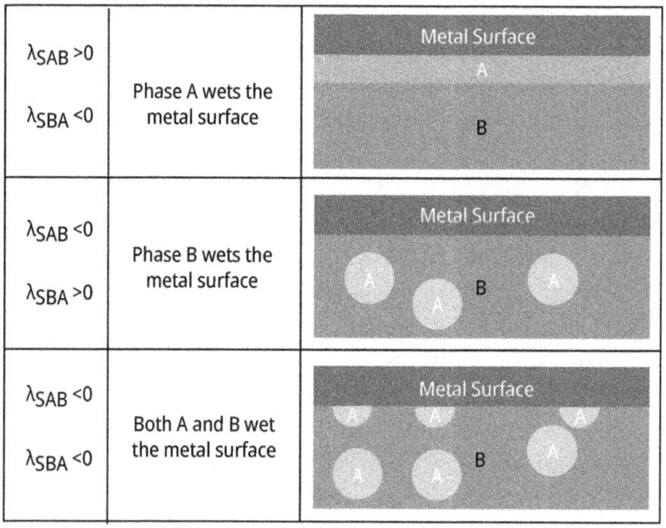

$\lambda_{SAB} > 0$ $\lambda_{SBA} < 0$	Phase A wets the metal surface	Metal Surface A B
$\lambda_{SAB} < 0$ $\lambda_{SBA} > 0$	Phase B wets the metal surface	Metal Surface A B A A
$\lambda_{SAB} < 0$ $\lambda_{SBA} < 0$	Both A and B wet the metal surface	Metal Surface A A A B A A A

Figure 7.7: Different surface morphologies of polymer blends in contact with a solid surface based on Harkin's equation.

Considering that the inner surfaces of extrusion dies are typically made of hydrophilic high surface energy stainless steel, it is likely to expect a higher level of interactions between the surface and the phase with higher polarity. Therefore, from the thermodynamic stand point, in a two-phase system with different polarities of components, it is likely to expect that the phase with higher polarity wets the surface of the die and, consequently, covers the surface of the final film. However, considering only thermodynamics cannot provide a reliable prediction for the localization of the phases at the film surface. It has been shown that in a two-phase flow, when the viscosity ratio of phases is far from unity, the phase with lower viscosity migrates and localizes itself close to the wall where higher shear rates exist [33]. In analyzing the final surface morphology of a polymer blend film, both thermodynamics and rheological properties as well as flow conditions (temperature and flow rate) should be considered. Therefore, it is likely to conclude that in a low interfacial tension system where polarity of the phases is similar, the viscosity ratio can play a much important role in determining the surface morphology while in a high interfacial tension system, the phase with higher polarity is more likely to wet the surface. Therefore, a low interfacial tension blend with viscosity ratio close to unity should be considered if the presence of both phases at the surface is desired.

7.4 Sealants based on immiscible polymer blends

The addition of a polymer with low seal initiation temperature (S.I.T.) to another polymer with higher S.I.T. is a common approach for reducing S.I.T. of a sealant film [34]. In the case of immiscible mixtures, low S.I.T. polymer should be located at the surface of the sealant film. Figure 7.8 schematically shows the arrangement of low S.I.T. phase on the surface in the seal area.

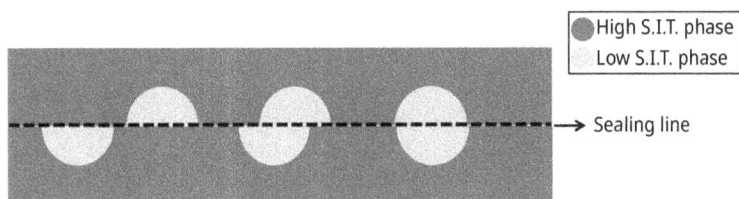

Figure 7.8: Seal area between two polymer blend films with low S.I.T. dispersed phase.

At temperatures above S.I.T. of the dispersed phase (but below S.I.T. of the matrix), only the dispersed phase domain at the surface can be melted and fused together and develop seal strength. At a low content of the dispersed phase, the effect is not considerable due to the small number of formed bridges between dispersed phases. By increasing the dispersed phase content, the effect becomes much pronounced as more bridges can be formed. In immiscible polymer blends where the S.I.T. of the two phases is at least 20° apart and both phases exist at the surface, the seal remains peelable between SIT of the phases. By increasing temperature above the S.I.T. of the matrix, matrix regions can also fuse to each other and contribute to the seal strength development. As a result, an increase in the seal strength is observed above the S.I.T. of the matrix. Mehta and Chen [35] showed that the addition of a very low density polyethylene (vLDPE)copolymer to polypropylene (PP) reduced its S.I.T. and broadened its peelable window (Figure 7.9). Table 7.1 lists common low SIT components that are used to reduce sealing temperature of PE-based sealants. The molecular structure and properties of these materials were discussed in detail in Chapter 4.

As another example, Figure 7.10 shows the effect of addition of 15 and 25 wt% of ethylene vinyl acetate (EVA) on seal strength of PE-based sealant. As can be seen, increasing EVA content from 0% to 25% reduced the S.I.T. of the sealant from 100 to 85 °C. Moreover, the addition of EVA broadened the seal curve.

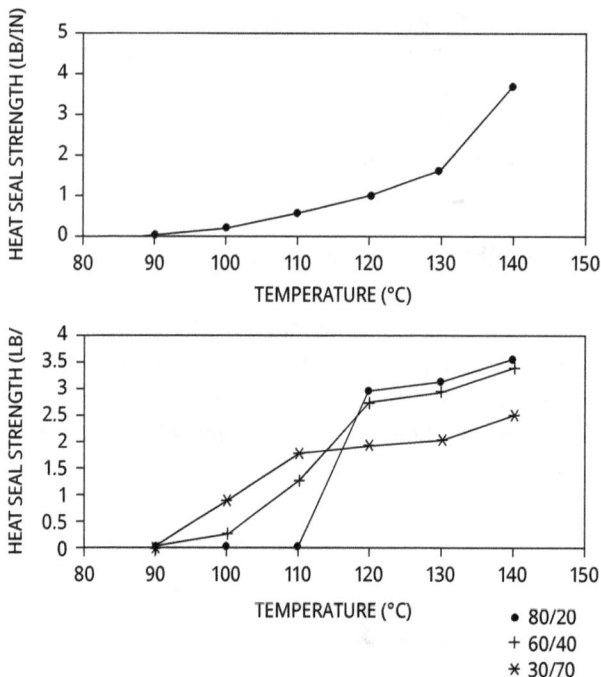

Figure 7.9: Seal strength of very low density polyethylene (vLDPE) and polypropylene blends at different blend ratios. The first two digits in the legend show vLDPE content while the second two digits show PP content [35].

Table 7.1: Common low S.I.T. polymers used in PE-based sealant films.

Low S.I.T. component	Peak melting temperature (°C)
Alpha olefin copolymers (plastomers)	90–100
Ethylene vinyl acetate (EVA)	47–100
Ethylene methyl-acrylate (EMA)	75–85
Ethylene-methacrylic acid (EMAA)	100–110
Ethylene-acrylic acid (EAA)	90–100
Ionomer	90–100
Ethylene *n*-butyl acrylate (EnBA)	60–70

Addition of polybutene-1 (PB-1) to polyethylene is another common strategy to broaden their peelable seal window [36–40]. Typically PB-1 compositions between 5 and 30 wt% are added to polyethylene for achieving peelable films [41, 42]. Nase et al. [43] showed that, in a blend of LDPE and PB-1, the peel force decreased exponentially by increasing the PB-1 content. The morphology of PB-1 in blends with polyethylene has been studied previously in literature and it was found that PB-1 forms a dispersed elongated phase in a polyethylene matrix [43, 44]. In addition, it has been shown that

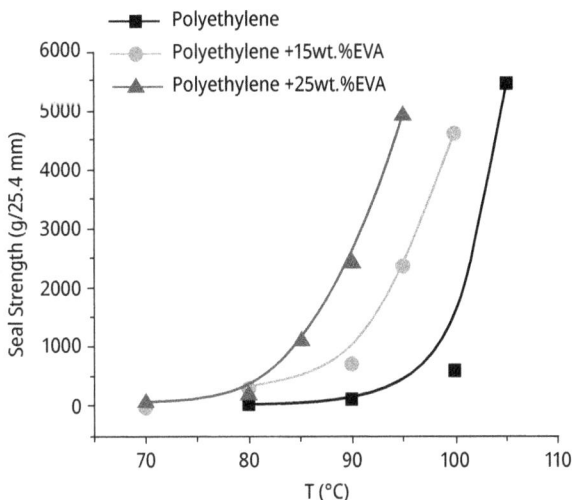

Figure 7.10: Effect of the addition of EVA to polyethylene on film seal strength.

cohesive failure within the sealant layer and along the interface of PE and dispersed PB-1 islands is the peeling mechanism in peelable sealants based on PB-1 [44]. The addition of polyolefin thermoplastic elastomers such as EPDM, EPM, halogenated butyl rubber, isoprene rubber, and styrene–butadiene rubber to polyolefin sealants has also been shown to result in a peelable sealant that can withstand high postprocessing temperatures such as retort applications [42]. The main disadvantage of using blend technology for peelable sealant is the fact that peeling properties is directly controlled with the blend morphology which is affected by the processing condition. Therefore, the same blend composition can result in different peeling forces when the film is processed at different conditions. This increases the risk of using this technology especially in the applications where consistent peel force is needed.

Another application for blending in sealant technology is the modification of the rheological properties of the sealant layer to enhance its caulkability. It has been shown that the caulkability of sealant materials increases considerably by reducing their S.I.T. and viscosity [45, 46]. Blending low S.I.T. and low viscosity sealants such as plastomers, ionomers, and metallocene polyethylene is typically considered as an effective approach where caulkability is needed. Forming miscible polymer blends is the goal in this approach to achieve viscosity reduction at low shear rates where caulking occurs [48].

7.5 Polymer nancomoposites

Polymer nanocomposites have received much attention recently due to their significant potential in altering different polymer blend properties such as their electrical [49], rheological [50, 51], and mechanical properties [52]. These materials have also received significant attention in polymer packaging due to their unique properties such as gas barrier, gas absorption, antimicrobial effects, sensing properties, and improved mechanical properties [53,54]. Previous studies on the addition of nanoparticles to sealant materials showed the potential of this approach in altering seal properties. Mohammadi et al. [62] compared seal properties of PE with organically modified montmorillonite (oMMT) and nonmodified MMT. They showed that while PE/oMMT nanocomposite exhibited a cohesive peel behavior with a peelable window of about 10 °C, PE film-containing MMT showed a lock seal performance. They also found a much lower peel strength with a significantly broad peelable window of 45 °C when PE-grafted-maleic anhydride (PE-g-MA) was used as a compatibilizer in PE nanocomposite. This was attributed to the fine distribution of oMMT in the nanocomposite with PE-g-MA that results in much more zones to form crack initiation during peeling. This mechanism is schematically shown in Figure 7.11.

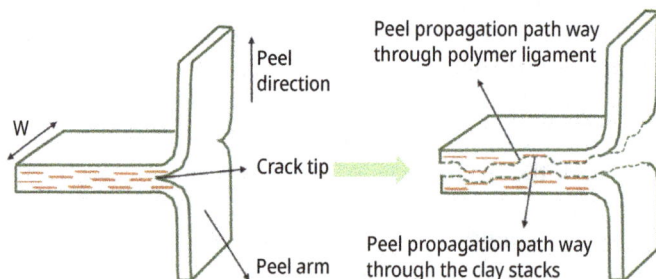

Figure 7.11: Peeling mechanism of PE/oMMT nanocomposite [61].

The addition of nanoparticles to polymer blends is another new approach in achieving high performance hybrid systems. Previous studies have shown that controlling the localization of solid particles in polymer blend is a key parameter in achieving desired properties. The thermodynamic equilibrium localization of solid particles in a polymer blend can be predicted by the Young's model [9]:

$$\omega = \frac{\gamma_{1s} - \gamma_{2s}}{\gamma_{12}} \tag{7.9}$$

where ω, γ_{1s}, γ_{2s}, and γ_{12} are the wetting parameter, the interfacial tensions between polymer 1 and solid, polymer 2 and solid, and polymer 1 and 2. If ω in equation (7.9) is greater than 1, then the localization of solid particles in phase 2 is thermodynamically preferred while for $\omega < -1$, the thermodynamic equilibrium localization of solid particles should be in phase 1. When $-1 < \omega < 1$, the localization of solid particles at the interface is thermodynamically preferred. The localization of solid particles is also affected by kinetic parameters. Previous studies showed that migration of solid particles from one phase to another in polymer blends can be considered as a three-step process: (i) migration from the bulk toward the interface, (ii) draining of matrix film between the particle and the interface, and (iii) migration at the interface. These steps are schematically shown in Figure 7.12.

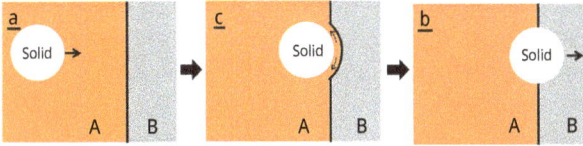

Figure 7.12: Different steps in migration of a solid particle from polymer A to polymer B: (a) bulk migration within polymer A, (b) film draining of polymer A, and (c) migration at the interface.

In the first step of migration, the particle moves within the initial polymer phase (polymer A in Figure 7.12). Eckstein et al. [56] studied experimentally the bulk migration and found that the particle flux in shear-induced migration scales as $(\dot{\gamma})$ where $(\dot{\gamma})$ is shear rate. The transitional velocity of spherical solid particles in the direction of the flow (flow-induced mechanism) has been shown to scale as $\dot{\gamma} \times R$ [57, 58]. These results indicate that both shear rate and particle size are important in the bulk migration step. In the film draining step, as the particle approaches the interface, a thin layer (film) of polymer A between the particle and the interface needs to be drained out so the particle can reach the interface. The film draining time (t_d) between a spherical solid particle with radius R and a deformable liquid/liquid interface can be estimated as

$$t_d = \frac{3n^2 \eta A_f^2}{16 \pi F_c} \left(\frac{1}{\delta_C^2} - \frac{1}{\delta_0^2} \right) \tag{7.10}$$

where F_c, η, A_f, δ_C, and δ_0 are the contact force, the viscosity of the polymer A, the surface area of the polymer A film between the particle and the interface ($\sim 2\pi R^2$), the critical film thickness in which the polymer A layer rupture occurs, and the initial polymer A film thickness where the film draining process begins, respectively. Moreover, n is the number of immobile interfaces in a system; in this case $n = 1$ as only the interface between the particle and polymer A is an immobile interface. In the particle migration theory, each particle has a limited time to drain the layer between itself and the interface which is called the contact time (t_c). If the draining time from eq. (7.10) is longer that the contact time, the particle will be moved away from the interface before it can enter the interface. Although estimating t_d and t_c are challenging, eq. (7.10) still can provide much insight into the effect of different processing conditions on particle migration. For example, increasing the shear rate increases the contact force but at the same time reduces the contact time between the particle and the interface [51]. In the third step, after the rupture of polymer A layer, the particle is located at the interface and begins migrating across the interface. Migration at the interface is a complex phenomenon in which interface and viscous forces compete. Jalali Dil and Favis [51] presented a semiquantitative model to predict the effect of different thermodynamic and kinetic parameters on particle migration at the interface.

Polymer hybrid systems where solid particles are added to polymer blends have also been used as a new approach in sealant technology. Previous studies showed that the addition of oMMT nanoparticles to PE/EVA blends increased the peelable window of the blend from 5 to 30 °C [59, 60]. Mohammadi et al. [61] studied the effect of the addition of oMMT on seal properties of polyethylene methacrylate copolymer (EMA), ethylene acrylic ester-glycidyl methacrylate terpolymer (EMA–GMA), and their blends with polyethylene. Using the Young's equation (eq. (7.9)) and by estimating interfacial tensions between the components, they found that oMMT nanoparticles should be located at the interface of PE/EMA and within EMA–GMA phase in PE/EMA–GMA. Using transmission electron microscopy and rheological data, they showed that oMMT nanoparticles were located in the predicted localizations.

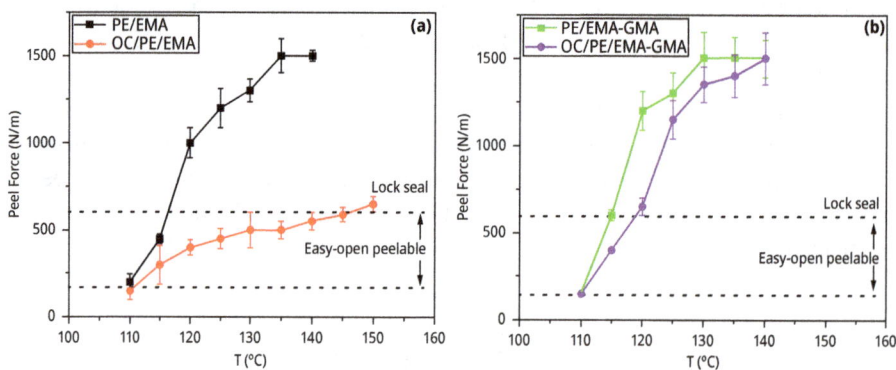

Figure 7.13: Seal curve of PE/EMA and PE/EMA--GMA blends and their nanocomposites with oMMT. The dashed lines show the peelable region [61].

When nanoparticles were located at the interface of PE/EMA, a significant change in the seal behavior was observed and peelable window was increased from 5 to 35 °C, Figure 7.13. On the other hand, the addition of oMMT to PE/EMA–GMA did not change considerably the seal behavior. This was attributed to initiation and growth of crack at the interface under stress due to the rigid nature of the nanoparticles while localization within EMA–GMA did not alter the interface crack phenomenon. Figure 7.14 schematically shows these two mechanisms.

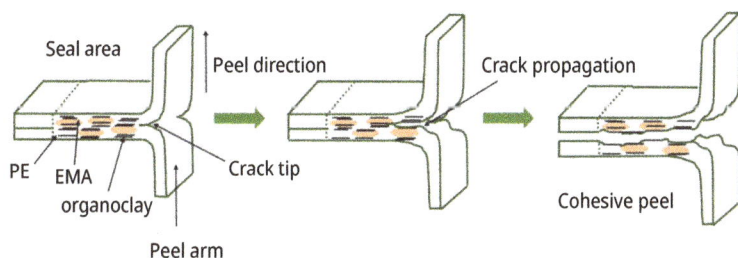

Figure 7.14: The effect of the interface localization of oMMT at the interface of PE/EMA on broadening of its peelable window.

It is worth mentioning that addition of nanoparticles and controlling their localization can also improve other properties of polymer blend-based packaging, such as their gas barrier properties. This indicates the multidimensional aspect of this approach compared to only blending approach and indicates the significant potential of this approach in improving different aspects of sealant films

References

[1]   Sperling, L.H., *Introduction to Physical Polymer Science*. 4th ed. 2006, New Jersey: John Wiley & Sons, Inc.

[2]   Flory, P.J., *Thermodynamics of High Polymer Solutions*. The Journal of Chemical Physics, 1942. **10**(1): p. 51–61.

[3]   Huggins, M.L., *Theory of Solutions of High Polymers1*. Journal of the American Chemical Society, 1942. **64**(7): p. 1712–1719.

[4]   BATES, F.S., *Polymer-Polymer Phase Behavior*. Science, 1991. **251**(4996): p. 898–905.

[5]   Coleman, M., *Specific Interactions and the Miscibility of Polymer Blends*. 2017, Routledge.

[6]   Wu, S., *Polymer Interface and Adhesion*. 1982.

[7]   Wu, S., *Interfacial and Surface Tensions of Polymers*. Journal of Macromolecular Science, Part C, 1974. **10**(1): p. 1–73.

[8]   Wu, S., *Calculation of interfacial tension in polymer systems*. Journal of Polymer Science Part C: Polymer Symposia, 1971. **34**(1): p. 19–30.

[9]   Elemans, P., J. Janssen, and H. Meijer, *The measurement of interfacial tension in polymer/polymer systems: The breaking thread method*. Journal of Rheology (1978-present), 1990. **34**(8): p. 1311–1325.

[10]   Carriere, C.J., and A. Cohen, *Evaluation of the interfacial tension between high molecular weight polycarbonate and PMMA resins with the imbedded fiber retraction technique*. Journal of Rheology, 1991. **35**(2): p. 205–212.

[11]   Carriere, C.J., A. Cohen, and C.B. Arends, *Estimation of interfacial tension using shape evolution of short fibers*. Journal of Rheology, 1989. **33**(5): p. 681–689.

[12]   Demarquette, N.R., and M.R. Kamal, *Interfacial tension in polymer melts. I: An improved pendant drop apparatus*. Polymer Engineering & Science, 1994. **34**(24): p. 1823–1833.

[13]   Palierne, J.F., *Linear rheology of viscoelastic emulsions with interfacial tension*. Rheologica Acta, 1990. **29**(3): p. 204–214.

[14]   Lacroix, C., M. Aressy, and P. Carreau, *Linear viscoelastic behavior of molten polymer blends: A comparative study of the Palierne and Lee and Park models*. Rheologica Acta, 1997. **36**(4): p. 416–428.

[15] Macosko, C.W., *Morphology development and control in immiscible polymer blends*. Macromolecular Symposia, 2000. **149**(1): p. 171–184.

[16] Dickie, R.A., *Mechanical Properties Small Deformations of Multiphase Polymer Blends, in Polymer Blends*, D.R. Paul, Editor. 2012, Elsevier Science.

[17] Favis, B.D., Factors influencing the morphology of immiscible polymer blends in melt ptocessing, in *Polymer Blends*, D.R. Paul, and C.B. Bucknall, Editors. 1999, Wiley: New York.

[18] Utracki, L., and Z. Shi, *Development of polymer blend morphology during compounding in a twin-screw extruder. Part I: Droplet dispersion and coalescence – A review*. Polymer Engineering & Science, 1992. **32**(24): p. 1824–1833.

[19] Li, J., P.L. Ma, and B.D. Favis, *The Role of the Blend Interface Type on Morphology in Cocontinuous Polymer Blends*. Macromolecules, 2002. **35**(6): p. 2005–2016.

[20] Favis, B.D., and J.-P. Chalifoux, *The effect of viscosity ratio on the morphology of polypropylene/ polycarbonate blends during processing*. Polymer Engineering & Science, 1987. **27**(21): p. 1591–1600.

[21] Willis, J.M., and B.D. Favis, *Processing-morphology relationships of compatibilized polyolefin/polyamide blends. Part I: The effect of an Ionomer compatibilizer on blend morphology*. Polymer Engineering & Science, 1988. **28**(21): p. 1416–1426.

[22] Utracki, L.A., and Z.H. Shi, *Development of polymer blend morphology during compounding in a twin-screw extruder. Part I: Droplet dispersion and coalescence – A review*. Polymer Engineering & Science, 1992. **32**(24): p. 1824–1833.

[23] Taylor, G.I., *The Viscosity of a Fluid Containing Small Drops of Another Fluid*. Proceedings of the Royal Society of London A 1, 1932. **138**: p. 41.

[24] Yuan, Z., and B.D. Favis, *Coarsening of immiscible co-continuous blends during quiescent annealing*. AIChE Journal, 2005. **51**(1): p. 271–280.

[25] Vinckier, I., et al., *Droplet size evolution during coalescence in semiconcentrated model blends*. AIChE Journal, 1998. **44**(4): p. 951–958.

[26] Chesters, A.K., *Modelling of coalescence processes in fluid-liquid dispersions: A review of current understanding*. Chemical Engineering Research and Design, 1991. **69(A4)**: p. 259–270.

[27] Lyu, S.-P., F.S. Bates, and C.W. Macosko, *Coalescence in polymer blends during shearing*. AIChE Journal, 2000. **46**(2): p. 229–238.

[28] Vinckier, I., et al., *Droplet size evolution during coalescence in semiconcentrated model blends*. AIChE Journal, 1998. **44**(4): p. 951–958.

[29] Jalali Dil, E., R. Silverwood, and A. Ajji, *Seal Performance of Bioplastic Poly(lactic acid)/Poly(butylene Adipate Co Terephthalate) Blends, in Polymer Processing Society (PPS) Americas*. 2018, Boston.

[30] Rezaei Kolahchi, A., P.J. Carreau, and A. Ajji, *Surface Roughening of PET Films through Blend Phase Coarsening*. ACS Applied Materials & Interfaces, 2014. **6**(9): p. 6415–6424.

[31] Harkins, W.D. and A. Feldman, *Films. The spreading of liquids and the spreading coefficient*. Journal of the American Chemical Society, 1922. **44**(12): p. 2665–2685.

[32] Hobbs, S.Y., M.E.J. Dekkers, and V.H. Watkins, *Effect of interfacial forces on polymer blend morphologies*. Polymer, 1988. **29**(9): p. 1598–1602.

[33] Wang, J., A. Reyna-Valencia, and B.D. Favis, *Controlling the continuity and surface migration of conductive poly(ether-block-amide) in melt processed cast-film blends*. Polymer, 2018. **136**: p. 224–234.

[34] Najarzadeh, Z., R.Y. Tabasi, and A. Ajji, *Sealability and seal characteristics of PE/EVA and PLA/PCL blends*. International Polymer Processing, 2014. **29**(1): p. 95–102.

[35] Mehta, A.K. and M.C. Chen, *Heat Sealable Blend of Very Low Density Polyethylene or Plastomer with Polypropylene Based Polymers and Heat Sealable Film and Articles Made Thereof*. 1994, Google Patents.

[36] Hwo, C.C., *Polybutylene Blends as Easy Open Seal Coats for Flexible Packaging and Lidding*. Journal of Plastic Film & Sheeting, 1987. **3**(4): p. 245–260.

[37] Hwo, C.C., *Polymer Packaging Film and Sheet Capable of Forming Peelable Seals Made from Ethylenic and Butene-1 Polymers*. 1992, Google Patents.

[38] Hwo, C.C., *Packaging Film and Sheet Capable of Forming Peelable Seals with Good Optics*. 1987, Google Patents.

[39] Denzel, H., and O. Hahmann, *Heat Sealable Polybutene-1 Blends Containing Polypropylene or Ethylene Copolymer*, 1978, Google Patents.

[40] Beer, J.S., *Peel Seal Blend of 1-polybutylene, m-LLDPE and LDPE with High Hot Tack*. 2002, Google Patents.

[41] Nase, M., et al., *Structure of blown films of polyethylene/polybutene-1 blends*. Polymer Engineering & Science, 2010. **50**(2): p. 249–256.

[42] Carespodi, D.L., *Peelable Film Laminate*. 1988, Google Patents.

[43] Nase, M., B. Langer, and W. Grellmann, *Fracture mechanics on polyethylene/polybutene-1 peel films*. Polymer Testing, 2008. **27**(8): p. 1017–1025.

[44] Sängerlaub, S., et al., *Identification of polybutene-1 (PB-1) in easy peel polymer structures*. Polymer Testing, 2018. **65**: p. 142–149.

[45] Mesnil, P., et al. *Seal Through Contamination Performance of Metallocene Plastomers*. In *TAPPI Polymers, Laminations and Coatings Conference, Chicago, IL*, August 2000.

[46] Ward, D., and M. Li, Seal-through-contamination and "caulkability" an evaluation of sealants' ability to encapsulate contaminants in the seal area, in *Tappi. 2016. Exploring New Frontiers*. 2016: Texas. p. 132.

[47] Najarzadeh, Z., A. Ajji, and J.-B. Bruchet, *Interfacial self-adhesion of polyethylene blends: The role of long chain branching and extensional rheology*. Rheologica Acta, 2015. **54**(5): p. 377–389.

[48] Kanani Aghkand, Z., and A. Ajji, *Squeeze flow in multilayer polymeric films: Effect of material characteristics and process conditions*. Journal of Applied Polymer Science, 2022. **139**(13): p. 51852.

[49] Jalali Dil, E., et al., *Interface Bridging of Multiwalled Carbon Nanotubes in Polylactic Acid/Poly(butylene adipate-co-terephthalate): Morphology, Rheology, and Electrical Conductivity*. Macromolecules, 2020. **53**(22): p. 10267–10277.

[50] Jalali Dil, E., and B.D. Favis, *Localization of micro and nano-silica particles in a high interfacial tension poly(lactic acid)/low density polyethylene system*. Polymer, 2015. **77**: p. 156–166.

[51] Jalali Dil, E., and B.D. Favis, *Localization of micro- and nano-silica particles in heterophase poly(lactic acid)/poly(butylene adipate-co-terephthalate) blends*. Polymer, 2015. **76**: p. 295–306.

[52] Jalali Dil, E., N. Virgilio, and B.D. Favis, *The effect of the interfacial assembly of nano-silica in poly (lactic acid)/poly (butylene adipate-co-terephthalate) blends on morphology, rheology and mechanical properties*. European Polymer Journal, 2016. **85**: p. 635–646.

[53] Youssef, A.M., *Polymer nanocomposites as a new trend for packaging applications*. Polymer-Plastics Technology And Engineering, 2013. **52**(7): p. 635–660.

[54] Taherimehr, M., et al., *Trends and challenges of biopolymer-based nanocomposites in food packaging*. Comprehensive Reviews in Food Science and Food Safety, 2021. **20**(6): p. 5321–5344.

[55] Fenouillot, F., P. Cassagnau, and J.C. Majesté, *Uneven distribution of nanoparticles in immiscible fluids: Morphology development in polymer blends*. Polymer, 2009. **50**(6): p. 1333–1350.

[56] Eckstein, E.C., D.G. Bailey, and A.H. Shapiro, *Self-diffusion of particles in shear flow of a suspension*. Journal of Fluid Mechanics, 1977. **79**(01): p. 191–208.

[57] Buyevich, Y.A., Fluid dynamics of fine suspension flow, in *Advances in the Flow and Rheology of Non-Newtonian Fluids, Part B*, D.A. Siginer, D. De Kee, and R.P. Chhabra, Editors. 1999, Elsevier: Netherlands. p. 1267.

[58] Bossis, G., and J.F. Brady, *Dynamic simulation of sheared suspensions. I. General method*. The Journal of Chemical Physics, 1984. **80**(10): p. 5141–5154.

[59] Manias, E., et al., *Polyethylene Nanocomposite Heat-Sealants with a Versatile Peelable Character*. Macromolecular Rapid Communications, 2009. **30**(1): p. 17–23.

[60] Zhang, J., et al., *Tailored Polyethylene Nanocomposite Sealants: Broad-Range Peelable Heat-Seals Through Designed Filler/Polymer Interfaces*. Journal of Adhesion Science and Technology, 2009. **23**(5): p. 709–737.

[61] Mohammadi, R.S., S.H. Tabatabaei, and A. Ajji, *Effect of nanoclay localization on the peel performance of PE based blend nanocomposite sealants*. Applied Clay Science, 2018. **152**: p. 113–123.

[62] Mohammadi, R.S., S.H. Tabatabaei, and A. Ajji, *Peelable clay/PE nanocomposite seals with ultra-wide peelable heat seal temperature window*. Applied Clay Science, 2018. **158**: p. 132–142.

Chapter 8
Bioplastic sealants

Commodity polymers are produced from nonrenewable petroleum resources and are not biodegradable or compostable. Considering the global capacity of 360 million metric tons for synthetic polymers, Shinoka et al. [1] indicated their significant environmental impact. Recently, polymers produced from renewable resources and biodegradable or compostable polymers have received an increasing amount of attention [2] as solutions for plastic environmental crisis. Bioplastics can be defined as polymers that are produced from bio-based resources and/or are biodegradable or compostable [3, 4]. This definition can be better understood using the schematic in Figure 8.1.

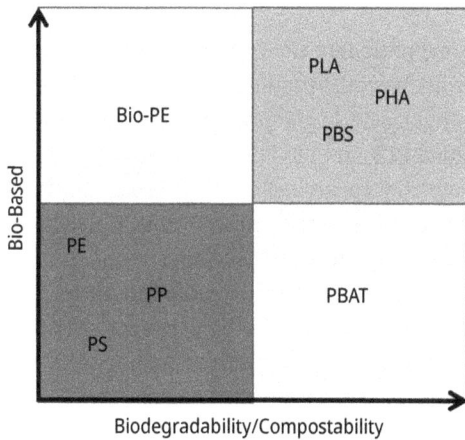

Figure 8.1: Classification of bioplastic materials, except the dark gray zone, other polymers are considered as bioplastics.

For example, polyethylene (PE) is a nonbiodegradable material but PE produced from bio-based resources is considered as a part of bioplastic family. This is due to the fact that bio-based PE is produced from a renewable resource (corn in this case) rather than nonrenewable petroleum resources.

Bioplastic family also has polymers such as poly(butylene adipate-co-terephthalate) or PBAT that are petroleum-based but biodegradable. Therefore, it can be seen that the definition of bioplastics is very broad. Despite such a broad definition of bioplastics, the majority of polymer materials do not fall within the bioplastic categories.

This has encouraged researchers in industry and academia to focus more on bioplastic materials. Despite their environmental advantages, bioplastic materials have not been able to play a major role in plastic industry. Based on the growth rate of bioplastics in 2010–2012, the annual bioplastic global capacity was expected to reach 6 million

https://doi.org/10.1515/9781501524592-008

tons at the end of 2017 [5] but the global production capacity of bioplastics in 2017 was determined as 2.1 million metric tons [6, 7]. This indicates a significantly lower growth rate than expected in the past decade. Despite the fact that bioplastics were initially presented as candidates for replacing commodity polymers, their poor mechanical properties limited significantly their applications in many applications. In addition, bioplastic materials are more expensive than commodity polymers. Despite the fact that bioplastics could not yet be a key player in plastic market, they still can be an interesting option to reduce environmental impact of plastic materials in some specific applications. In this chapter, some bioplastic polymers that are used in flexible packaging are reviewed [5].

8.1 Bio-based polyethylene terephthalate (bio-PET)

The molecular structure of polyethylene terephthalate (PET) and its characteristics were discussed in detail in Chapter 4. PET is industrially produced by condensation polymerization of ethylene glycol and terephthalic acid in the presence of a catalyst [8]. The main difference between the conventional PET and bio-PET is that ethylene glycol used in the production of bio-PET is obtained from sugarcane or sugar beets. Considering the molecular structure of PET, ethylene glycol comprises 30 wt% of PET, which indicates that bio-PET is a 30 wt% bio-based material. Although there have been attempts from giant beverage companies to commercialize 100% plant-based bio-PET since 2011, it has not yet reached the commercialization stage and is still in the prototype step. Bio-PET has been the major player of bioplastics especially in rigid packaging with the global production capacity of about 25% of global bioplastic production capacity.

Recently and following the announcement of the production of bio-based paraxylene (a key raw material for production of terephthalic acid) from beet sugar by Virent Inc. [9], producing 100% bio-based PET has become possible. It should be noted that similar to fossil-based PET, bio-PET is not biodegradable/compostable but can be recycled to reduce its environmental impact.

8.2 Bio-based Polyethylene (PE) and Ethylene Vinyl Acetate (EVA)

Braskem in Brazil began production of bio-based polyethylene in 2010. The main difference of bio-based polyethylene and petroleum-based polyethylene is that in bio-PE production, ethylene is produced from sugarcane rather than from petroleum resources. Currently, Braskem produced bio-LDPE, bio-LLDPE, and bio-HDPE. Their LLDPE portfolio includes both conventional butane and hexane copolymers. It should be noted that the highest bio-based content in their portfolio is their bio-HDPE with 96% bio-based material. The properties of bio-based PE are similar to petroleum-based ones but its higher price and limited availability have limited its application. Braskem recently began producing bio-based EVA with two grades having 45% and 80% bio-based content.

8.3 Poly(lactide)

Among the family of biodegradable/compostable polyesters, polylactide (PLA) has been the focus of much attention due to the following reasons: (i) it is produced from renewable resources such as corn, (ii) is compostable; (iii) has very low or no toxicity, (iv) has high mechanical stiffness and clarity, and (v) good availability in the market [10]. Figure 8.2 shows different molecular structures of PLA. Properties of PLA are highly related to the ratio between the two isomers D and L in the PLA molecular structure. Commercially available, we can find 100% L-PLA (called PLLA) with high level of crystallinity and copolymers of poly(L-lactic acid) and poly(D,L-lactic acid) which are rather amorphous. Among compostable bioplastics, PLA has a high mechanical strength and clarity. This high mechanical strength could be disadvantageous in some application due to flex cracking of PLA films. Anderson et al. [11] reviewed different strategies to increase toughness of PLA including variation of mesoform ratio in the polymer, plasticization, and blending with other polymers. Another challenge with PLA is its difficult film production due to its low melt strength and high stiffness. Previous studies emphasized on the importance of optimizing processing conditions such as extrusion temperature and line speed to achieve a staple extrusion process [12]. Due to its high clarity and stiffness, PLA has been considered as a candidate to replace PET.

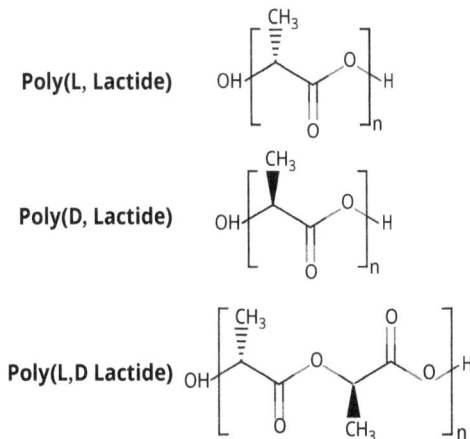

Figure 8.2: Different molecular structure of poly(lactide), PLA.

In 2015, Natureworks and Bi-AX International announced the development of biaxially oriented PLA film to replace BoPET. PLA can also be used as a sealant layer specially when laminated to paper or other compostable substrates to create fully compostable packaging. Sealing properties of PLA films have been investigated by some researchers. For example, Yousefzadeh and Ajji [13] examined sealing properties of an amorphous PDLA film and found a seal initiation temperature at around 80 °C.

8.4 Polycaprolactone

Poly(ε-caprolactone) or PCL is a polyester that is obtained by ring-opening polymerization of ε-caprolactone in the presence of a catalyst. Figure 8.3 shows the molecular structure of PCL.

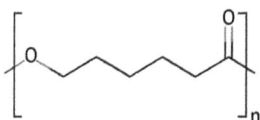

Figure 8.3: Molecular structure of PCL.

Biodegradable character of PCL has made it an interesting option for controlled release of drugs [10]. PCL has a low T_g (~−60 °C) and a low melting point (65–80 °C), which indicates that PCL has a low thermal resistance. Although the soft and biodegradable nature of PCL shows that it could be an interesting option for biodegradable flexible packaging [14], its low melting temperature has limited its applications. Therefore, PCL is generally blended or modified (e.g., copolymerization, cross-linking) [15]. For example, Li and Favis [16] studied the morphology and miscibility of PCL blends with thermoplastic starch (TPS) as a route to produce higher thermal resistance bioplastic blend. They found that PCL has the tendency to encapsulate TPS in this blend. In addition, they found the presence of strong hydrogen bonding interactions between the carbonyl groups of the PCL and the hydroxyl groups on the starch. Tabasi and Ajji [17, 18] examined using PLA/PCL blends for flexible film application and found that the addition of PCL reduced hot tack initiation temperature and increased hot tack plateau seal strength (Figure 8.4).

8.5 Aliphatic polyesters and copolyesters

A large number of aliphatic polyesters and copolyesters are biodegradable polymers that are produced from petroleum resources. Mochizuki [19] reviewed properties and applications of aliphatic polyesters in detail. Among them, polybutylene succinate (PBS) and copolyester of poly(butylene succinate/adipate), PBSA, received much attention. Some previous studies have shown interesting properties for PBS and PBSA such as

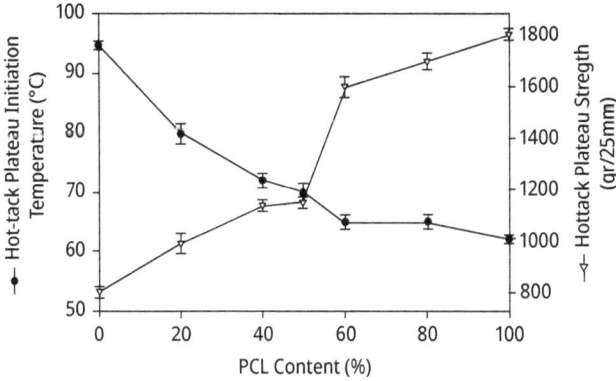

Figure 8.4: The effect of addition of PCL on the hot tack properties of PLA/PCL blends [17].

higher oxygen and water vapor moisture [20]. Figure 8.5 shows the molecular structures of PBS and PBSA.

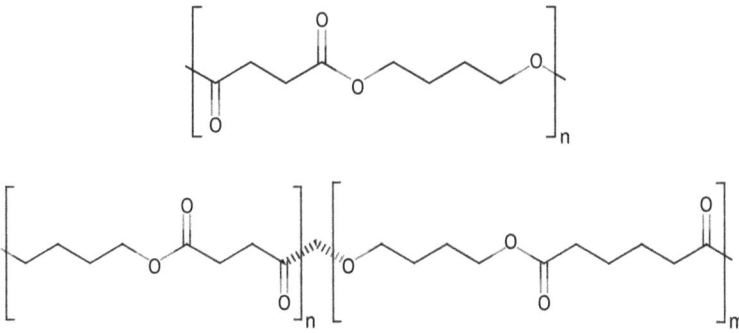

Figure 8.5: Molecular structure of PBS (top) and PBSA (bottom).

Due to its copolymer nature, PBSA properties can be tailored by changing the ratio of comonomers. For example, its melting temperature can vary from 115 °C (almost the same as PBS) to below room temperature by increasing butylene adipate content [21]. Previous studies also showed that increasing butylene adipate content in PBSA increases its biodegradation rate [22]. Lai et al. [23] studied the blending of PBS with TPS and showed that adding glycerol as a compatibilizer/plasticizer is necessary in achieving good dispersion. They also found a considerable increase in the crystallinity of PBS with the addition of TPS as it played a nucleating role for PBS. Blends of PBS and PLA have been shown to form an immiscible polymer blend [24, 25]. Despite its much better gas barrier, the milky look of PBS films has limited its application especially in many food packaging where transparency is needed. In addition, due to its high degradation rate, cautions have to be made when using PBS or PBSA as a sealant in contact with foods that contain active bacteria such as yeasts as it can accelerate its degradation and result in packaging failure [26].

8.6 Aromatic copolyesters

Compared to totally aliphatic copolyesters, aromatic copolyesters are often based on terephthalic diacid. Among them, PBAT has received much attention. Figure 8.6 shows the molecular structure of PBAT. It is a random copolymer of butylene adipate and butylene terephthalate.

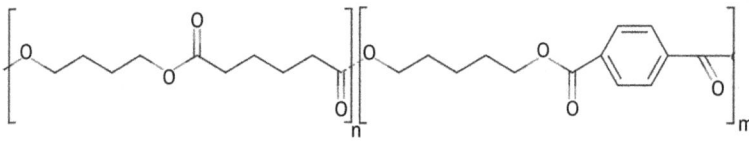

Figure 8.6: Molecular structure of PBAT: left sequence shows butylene adipate and the right segment shows butylene terephthalate.

BASF commercialized PBAT with Ecoflex® trademark which is still the most known trademark in the market. Aromatic copolyesters, like aliphatic copolyesters, degrade completely in the microorganism environment. An increase of terephthalic acid content tends to decrease the degradation rate [10]. Ecoflex biodegradation has been analyzed by Witt et al. [27] but they could not find any indication for environmental risk (ecotoxicity) of the composting residue of PBAT. A detailed review on Ecoflex properties and processing was presented by Yamamoto et al. [28]. The low stiffness of PBAT has limited its use in some applications. Previous studies showed the potential of blending PBAT with PLA to achieve a balance between stiffness and toughness [29, 30]. Table 8.1 shows the effect of different PBAT contents on properties of PLA/PBAT blends.

Table 8.1: Effect of the addition of PBAT on mechanical properties of PLA [29].

Sample composition	Tensile modulus (MPa)	Yield strength (MPa)	Elongation at break (%)	Impact strength (J/m)
PLA	2,130	62.5	4.2	22
PLA/PBAT(80/20)	1,720	53	122	31
PLA/PBAT(70/30)	1,540	46.7	183	55
PLA/PBAT(50/50)	965	28	315	275

PLA/PBAT blend was patented and commercialized under the trademark of Eco-vio by BASF [31]. It has been found that the addition of chain extenders with epoxy functional group improves processability and final mechanical properties in these blends. Jalali et al. used different analysis techniques and showed that the PLA/PBAT blend shows a complex partial miscible behavior [32]. Tabasi and Ajji examined the heat seal properties of PLA/PBAT blends and showed that the addition of PBAT to PLA is an effective approach in reducing the seal initiation temperature of the blend [13]. Blending of PBAT with TPS is another approach to improve its stiffness. This blend was commercialized under the trademark of Mater-Bi by Novamont [33].

8.7 Polyhydroxyalkanoates (PHA)

Polyhydroxyalkanoate (PHA) is a family of bioplastics produced from microbial fermentation of carbon feed stock. Poly(hydroxybutyrate) (PHB) and poly(3-hydroxybutyrate-co-3-hydroxyvalerate) (PHBV) are the most known members of the PHA family. Figure 8.7 shows the molecular structures of PHB and PHBV.

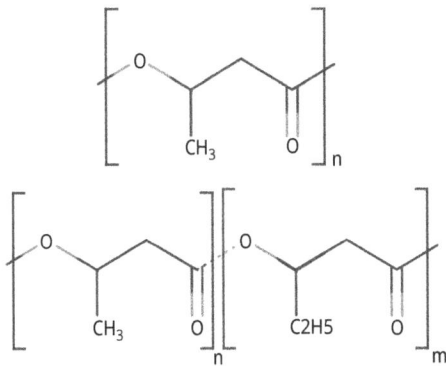

Figure 8.7: Molecular structures of PHB (top) and PHBV (bottom).

PHB has high crystallinity and high melting point (~180 °C), which makes it an interesting candidate to replace PP for high-temperature applications such as hot beverage or retort applications. However, its brittle behavior and narrow processing window make the use of PHB very challenging in flexible packaging [34]. PHBV has lower crystallinity and melting temperature (~145 °C), which allows easier processing of the resin; however, it does not provide the same thermal resistance as of PHB.

References

[1] Shinoka, T., et al., *Creation of viable pulmonary artery autografts through tissue engineering.* The Journal of Thoracic and Cardiovascular Surgery, 1998. **115**(3): p. 536–546.

[2] Yu, L., K. Dean, and L. Li, *Polymer blends and composites from renewable resources.* Progress in Polymer Science, 2006. **31**(6): p. 576–602.

[3] Martin, O. and L. Avérous, *Poly(lactic acid): Plasticization and properties of biodegradable multiphase systems.* Polymer, 2001. **42**(14): p. 6209–6219.

[4] Avérous, L. and E. Pollet, *Biodegradable Polymers, in Environmental Silicate Nano-Biocomposites,* L. Avérous and E. Pollet, Editors. 2012, Springer London. p. 13–39.

[5] European-Bioplastics, *Bioplastics Facts and Figures.* 2013, European Bioplastics.

[6] Lantz, G.C., et al., *Small Intestinal Submucosa as a Vascular Graft: A Review.* Journal of Investigative Surgery, 1993. **6**(3): p. 297–310.

[7] European-Bioplastics, *Bioplastics Facts and Figures.* 2019, European bioplastics.

[8] Barber, N.A., *Polyethylene Terephthalate: Uses, Properties and Degradation.* 2017, Nova Science Publishers.

[9] *Virent Bioformpx® Paraxylene used to produce World's First 100% plant based polyester shirts.* 2016 19 November 2019]; Available from: https://www.virent.com/news/virent-bioformpx-paraxylene-used-to-produce-worlds-first-100-plant-based-polyester-shirts/.

[10] Avérous, L., *Biodegradable multiphase systems based on plasticized starch: A review.* Journal of Macromolecular Science, Part C, 2004. **44**(3): p. 231–274.

[11] Anderson, K.S., K.M. Schreck, and M.A. Hillmyer, *Toughening polylactide.* Polymer Reviews, 2008. **48**(1): p. 85–108.

[12] Karkhanis, S.S., et al., *Blown film extrusion of poly(lactic acid) without melt strength enhancers.* Journal of Applied Polymer Science, 2017. **134**(34): p. 45212.

[13] Tabasi, R. and A. Ajji, *Tailoring heat-seal properties of biodegradable polymers through melt blending.* International Polymer Processing, 2017. **32**(5): p. 606–613.

[14] Lyu, J.S., J.-S. Lee, and J. Han, *Development of a biodegradable polycaprolactone film incorporated with an antimicrobial agent via an extrusion process.* Scientific Reports, 2019. **9**(1): p. 20236.

[15] Averous, L., *Biodegradable multiphase systems based on plasticized starch: A review.* Journal of Macromolecular Science, Part C: Polymer Reviews, 2004. **44**(3): p. 231–274.

[16] Li, G. and B.D. Favis, *Morphology development and interfacial interactions in polycaprolactone/thermoplastic-starch blends.* Macromolecular Chemistry and Physics, 2010. **211**(3): p. 321–333.

[17] Tabasi, R.Y., Z. Najarzadeh, and A. Ajji, *Development of high performance sealable films based on biodegradable/compostable blends.* Industrial Crops and Products, 2015. **72**: p. 206–213.

[18] Najarzadeh, Z., R.Y. Tabasi, and A. Ajji, *Sealability and seal characteristics of PE/EVA and PLA/PCL blends.* International Polymer Processing, 2014. **29**(1): p. 95–102.

[19] Mochizuki, M. and M. Hirami, *Structural effects on the biodegradation of aliphatic polyesters.* Polymers for Advanced Technologies, 1997. **8**(4): p. 203–209.

[20] de Matos Costa, A.R., et al., *Properties of biodegradable films based on poly (butylene succinate)(PBS) and poly (butylene adipate-co-terephthalate)(PBAT) blends.* Polymers, 2020. **12**(10): p. 2317.

[21] Aliotta, L., et al., *A brief review of Poly (Butylene Succinate)(PBS) and its main copolymers: Synthesis, blends, composites, biodegradability, and applications.* Polymers, 2022. **14**(4): p. 844.

[22] Ichikawa, Y. and T. Mizukoshi, *Bionolle (polybutylenesuccinate).* Synthetic Biodegradable Polymers, 2011: p. 285–313.

[23] Lai, S.M., C.K. Huang, and H.F. Shen, *Preparation and properties of biodegradable poly(butylene succinate)/starch blends.* Journal of Applied Polymer Science, 2005. **97**: p. 257–264.

[24] Zhao, P., et al., *Preparation, mechanical, and thermal properties of biodegradable polyesters/poly (lactic acid) blends.* Journal of Nanomaterials, 2010. 2010.

[25]   Shibata, M., Y. Inoue, and M. Miyoshi, *Mechanical properties, morphology, and crystallization behavior of blends of poly (L-lactide) with poly (butylene succinate-co-L-lactate) and poly (butylene succinate).* Polymer, 2006. **47**(10): p. 3557–3564.

[26]   Thirunavukarasu, K., et al., *Degradation of poly (butylene succinate) and poly (butylene succinate-co-butylene adipate) by a lipase from yeast Cryptococcus sp. grown on agro-industrial residues.* International Biodeterioration & Biodegradation, 2016. **110**: p. 99–107.

[27]   Witt, U., et al., *Biodegradation of aliphatic–aromatic copolyesters: Evaluation of the final biodegradability and ecotoxicological impact of degradation intermediates.* Chemosphere, 2001. **44**(2): p. 289–299.

[28]   Yamamoto, M., et al., *Biodegradable aliphatic-aromatic polyesters: "Ecoflex®",* in *biopolymers online.* 2005, Wiley-VCH Verlag GmbH & Co. KGaA.

[29]   Dil, E.J., N. Virgilio, and B.D. Favis, *The effect of the interfacial assembly of nano-silica in poly (lactic acid)/poly (butylene adipate-co-terephthalate) blends on morphology, rheology and mechanical properties.* European Polymer Journal, 2016. **85**: p. 635–646.

[30]   Jiang, L., B. Liu, and J.W. Zhang, *Properties of Poly(lactic acid)/Poly(butylene adipate-co-terephthalate) /nanoparticle ternary composites.* Industrial & Engineering Chemistry Research, 2009. **48**(16): p. 7594–7602.

[31]   Freese, F., et al., *Biodegradable Polyester Film.* 2014, Google Patents.

[32]   Jalali Dil, E., P.J. Carreau, and B.D. Favis, *Morphology, miscibility and continuity development in poly (lactic acid)/poly(butylene adipate-co-terephthalate) blends.* Polymer, 2015. **68**: p. 202–212.

[33]   Borchani, K.E., C. Carrot, and M. Jaziri, *Biocomposites of Alfa fibers dispersed in the Mater-Bi® type bioplastic: Morphology, mechanical and thermal properties.* Composites Part A: Applied Science and Manufacturing, 2015. **78**: p. 371–379.

[34]   Barham, P. and A. Keller, *The relationship between microstructure and mode of fracture in polyhydroxybutyrate.* Journal of Polymer Science Part B: Polymer Physics, 1986. **24**(1): p. 69–77.

Chapter 9
Case studies

In this chapter, we present some examples of sealant design for different flexible packaging applications. Some aspects that should be considered during designing their sealant film in particular are reviewed. Presented equations in this chapter are based on the following assumptions:

- No patterned jaw is used in the sealing.
- No seal imperfection exists in the package.
- The force that acts on the seal area during packaging is applied within a timescale comparable to the timescale of the force applied during the seal/hot tack test.

It should be noted that these assumptions are critical to be able to establish a relation between the seal test results and the required seal force in different applications. The equations presented in this chapter should be used to provide guidelines for sealant design and should be combined by considering different aspects related to the product, film, and packaging methods.

9.1 Design of a peelable sealant for cereal packaging

You were asked to design a peelable film for packaging of 700 g cereal in a pillow pouch with the dimensions of 35 × 25. Which of the following sealants can be used for this application?

Table 9.1: Seal and Hot tack strength of peelable sealant materials for dry food application.

Sealant material name	Seal strength (N/25 mm) at jaw temperature	Hot tack strength (N/25 mm) at jaw temperature
AX-01	5.5	2.2
AX-02	6.2	3.5
AX-03	10.8	7

9.1.1 Solution

Cereals are usually packaged on VFFS machines. The package filling process begins when the jaws are sealing the bottom seal, but the jaws are opened before the completion of the filling cycle. This exposes the sealed area to the product weight while the seal is still hot. This clearly indicates that hot tack strength of the sealant is important

https://doi.org/10.1515/9781501524592-009

during the packaging process of cereal and should be considered as the designing parameter.

The required hot tack strength can be estimated by considering the force balance around the seal area at the bottom of the package as visualized in Figure 9.1.

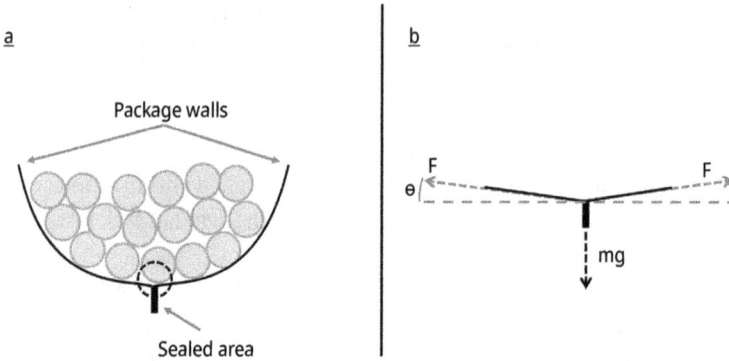

Figure 9.1: (a) Bottom seal area in a filled package and (b) higher magnification of the marked area around the sealed area in (a) with acting force vectors.

Using force balance at the seal area, we can obtain the following equation for the force that is applied to the seal area:

$$F = \frac{0.025 \times m.g}{2W \sin \theta} \tag{9.1}$$

where F is the required seal force, W is the package width, and θ is the angle between the force vectors (F) and the product weight (see Figure 9.1(b)). It is worth mentioning that the factor of 0.025 was used to obtain hot tack strength with the unit of N/25 mm. When the package is filled with the product, θ is usually small and challenging to determine. However, by experience, 10° is very common in many different pillow pouch and pinched bottom packaging. Using this assumption, equation (9.1) can be simplified as

$$F = \frac{0.072 \times m.g}{W} \tag{9.2}$$

The required hot tack strength for packaging of 700 g cereal can then be estimated as ~2 N/25 mm. To ensure seal integrity, it is always recommended to consider a safety factor in designing sealant materials. This means that selecting AX-01 is a risky choice but both AX-02 and AX-03 should provide enough hot tack to handle 700 g of cereal.

Cereal packaging is made of films containing high HDPE content to provide high moisture barrier needed for the product [1] as well as dead fold feature as a reclosure option. However, high HDPE content in these films leads to a low tear resistance in the bag length direction due to higher molecular orientation [2]. If the seal strength of the sealant is high, such as for AX-03, the risk of package tear and failure during open-ing by the final consumer is high. Therefore, among the presented options, selecting AX-02 should provide the best solution for this application.

9.2 Sealant for liquid packaging

A customer asked for a sealant film to package 1.5 L of milk on the VFFS machine. The package dimensions are 32.5 × 15 cm. Which of the following sealants will be your recommendation?

Sealant material name	Seal strength (N/25 mm) at jaw temperature	Hot tack strength (N/25 mm) at jaw temperature
BX-1	40 (b)	5
BX-2	12 (p)	7
BX-3	22 (p)	10
BX-4	40 (b)	12

*p, peelable; b, break.

9.2.1 Solution

Considering a density of ~1 kg/L for milk, the product weight can be estimated as ~1.5 kg. In the VFFS packaging of milk, the filling cycle begins when the jaws are closed but it continues even after jaw opening. This indicates that the seal will be exposed to the product weight while it is hot and therefore hot tack strength should be the first design parameter [3]. Using equation (9.2), the required hot tack strength to withstand 1.5 kg of milk can be estimated as 7.35 N/25 mm. Considering the available options and safety factor, BX-03 and BX-04 could be interesting options for the sealant.

For packaging of liquids, one should consider that the liquid can move in the package during transportation and final use which compresses the air in the package overhead space and can cause high stress on the seal area [4]. As a peelable feature does not add a value for this application (milk packaging) and just increases the risk of spillage, selecting BX-04 will be a safe option to ensure seal integrity during packag-ing, transportation, and handling.

9.3 Film for heavy-duty shipping sack (HDSS)

A customer would like to package 10 kg of salt in a pillow-shaped bag with dimensions of 46 cm × 30 cm. Which of the following films and what thickness can provide the required performance?

Sealant material name	Film yield strength (MPa)	Seal strength (N/25 mm) at jaw temperature	Price ($/lb)
CX-01	9	35 (b)	1
CX-02	13	35 (b)	1.2
CX-03	10	40 (b)	1.1

9.3.1 Solution

As heavy-duty sack (HDSS) bags are mostly premade (sealed and cut from the roll-stock and then filled with product in the next step) and their bottom is supported during and after the filling process, the filling process is not the step where the seal area is exposed to the product weight. Palletization, transportation, and handling are the steps where the seal needs to stay intact and protect the product in this application. Therefore, seal strength should be considered as the main sealant design parameter. Using equation (9.2), the minimum seal strength required to handle 10 kg of product in this bag can be estimated as 24 N/25 mm. In HDSS application, the stress in the film should not exceed its yield strength because of yielding and stretching of the film. In order to avoid film stretching, the stress in the film should always be kept below 60% of its yield strength. The stress in bag wall can then be estimated using the following equation which considers this safety factor:

$$\sigma = \frac{1.5 \times F \times W/0.025}{W \times b} = \frac{1.5 \times F}{0.025 \times t} \tag{9.3}$$

where b is the film thickness. Using $F = 24$ N/25 mm and yield strength of the films for σ, the required thickness of the films can be estimated as 160 μm, 111 μm, and 144 μm for CX-01, CX-02, and CX-03, respectively. By assuming a similar density for the films, these results indicate that the quantity of CX-02 film needed to produce a certain number of bags will be the lowest among the three films, and therefore, this will be the most economic and sustainable (as it reduces the amount of plastic use) option for the application.

9.4 Peelable film for over-the-mountain packaging

A customer is looking for a peelable film to package snacks. As it will be shipped to a destination at the altitude of 3,000 m, the package needs to provide hermetic seal for over-the-mountain (OTM) packaging. The package is a pillow pouch with the dimensions of 12.5 × 11 × 2.5 cm (Figure 9.2). What seal force do you recommend for the application? What would be the seal force if the customer decides to send the package by airplane?

9.4.1 Solution

In OTM applications, the package experiences high altitude or low outside pressures, and the bag inflates as is schematically shown in Figure 9.2. Therefore, the objective here is to evaluate the effect of pressure change on the seal integrity. If we consider the bag as a tube, then the largest equator (D in Figure 9.2(b)) can be estimated using the following equation:

$$D = \frac{2W}{\pi} \tag{9.4}$$

where W is the package width (Figure 9.2(a)). Using the package width of 11 cm, D can be estimated as 7 cm for this package.

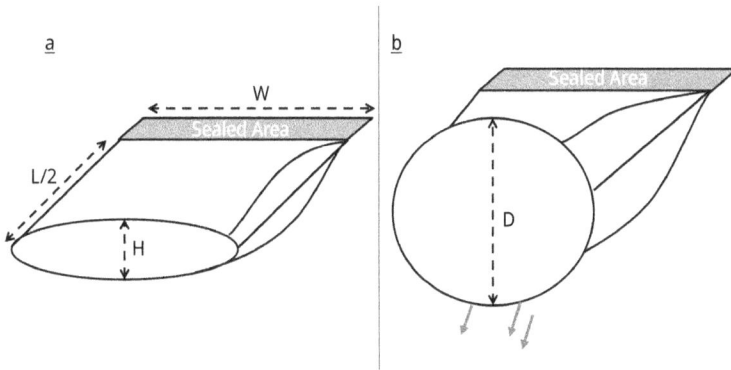

Figure 9.2: (a) The cross section of a pillow pouch at room pressure and (b) the cross section of the pouch in (a) inflated at high altitude due to the lower outside pressure.

Using the same concept as the hoop stress in the pipes [5] and by considering the force balance between the applied pressure force to the package cross-sectional area $\left(\mu D^2/4\right)$, and the force induced in the package wall (or the film), we can obtain an estimation formula for the force applied to the bottom and top-sealed area in the package due to the pressure difference between outside and inside of the package:

$$F = \frac{\Delta P \times \pi D^2}{8} \times \frac{0.025}{\pi D} = \frac{0.025 \times \Delta P \times D}{8} \tag{9.5}$$

where ΔP is the pressure difference between outside and inside of the package, and D is the largest equator of the cross section of the inflated bag determined from equation (9.3). It should be noted that $0.025/\pi D$ was multiplied to obtain seal strength with the unit of N/25 mm.

The air pressure at 3,000 m altitude can be estimated using the Laplace barometric formula [6] as 70.1 kPa. By assuming that the package was filled at the sea level, a pressure difference of 31.2 kPa between outside and inside of the package can be estimated at the destination. Using equation (9.6), the required seal strength can then be estimated as 6.8 N/25 mm. This should be considered as the minimum required peel force.

When the customer uses airplanes to ship, we should consider that nonpressurized feeder aircraft typically fly at approximately 4,500 m altitude. The air pressure at this altitude can be estimated as 57.7 kPa which leads to a pressure difference of $\Delta P = 43.6$ kPa during shipping. Using equation (9.5), we can estimate the minimum required seal force as 9.5 N/25 mm. These results clearly point to the importance of the knowledge of package transportation method when designing a package specially for light weight air-filled products such as snacks.

9.5 Sealant film with caulkability for spouted pouch application

You were asked to select a sealant material for 90 and 250 mL spouted pouches filled with apple puree that can provide caulkability around the fitment. Figure 9.3 schematically shows the pouch. They are made in a form fill seal machine where the three sides are sealed and then the pouch is filled from the top and finally capped spouts are inserted and sealed from the top. Figure 9.4 shows the viscosity curves of the

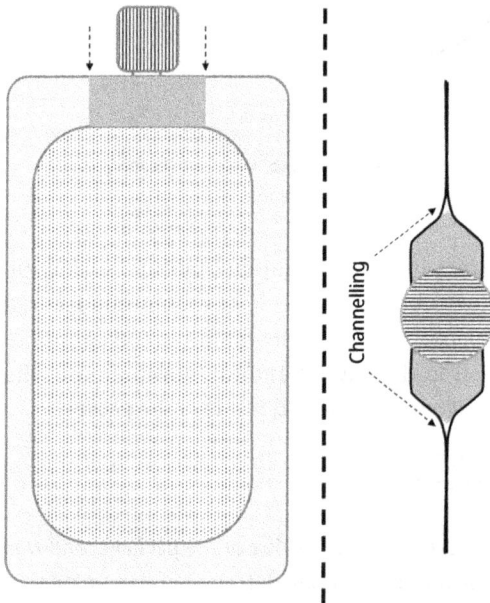

Figure 9.3: (Left) Spouted pouch: the spout is shown as the gray area at the top with the twist cap connected to it. Two arrows show spout corners. (Right) Top view of the spout pouch shown in left with arrows showing spout corners with channels between the films.

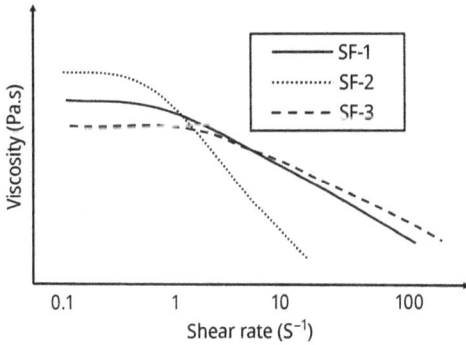

Figure 9.4: Viscosity curves of three sealants available for spouted pouch application.

available options for the sealant materials. Which sealant and at what thickness would you recommend for this application?

9.5.1 Solution

One of the major week points of spouted pouches is on the edge of the spout part of the fitment which are shown with arrows in Figure 9.3. In these areas, as the two sealants are joined at the corner of the spout, a high chance of channeling exists that can result in product spillage after filling. In order to ensure that the sealant material flows and fills the gap at the corners of the spout, a sealant with high caulkability or squeeze-out flow (SOF) is needed for this application. As discussed in Chapter 6, precise comparison of SOF between sealants requires simultaneous simulation of heat and squeeze flow. However, as was shown by Kanani Aghkand and Ajji [7], the shear rate of SOF in a wide range of sealing pressure falls between 1 and 10 s^{-1}. Using this rule of thumb, although SF-2 has a higher zero-shear viscosity, it should provide a better caulkability as it has the lowest viscosity in this shear rate range.

As was also shown in Chapter 6, in addition to sealant melt viscosity, increasing sealant thickness is a very efficient approach in enhancing SOF. Kanani Aghkand and Ajji [7] showed that while 50 μm sealants showed little SOF (1–2%) even at low viscosities, increasing sealant thickness to above 100 μm could improve significantly SOF (~25%). In spouted pouches, the size of the pouch indicates the size of the spout that should be used; the larger the pouch, the larger will be the spout and more precautions need to be taken when designing the sealant. For small product volumes such as 90 mL, standard spouts of 8.6 mm diameter are very common while for 250 mL pouches, 10 or even 16 mm spout are used. This indicates that while 50 μm sealant could be enough to provide reliable sealing around the spout in 90 mL pouches, a thicker sealant of 75 um should be used for 250 mL format to avoid channeling around the spout.

9.6 Peelable barrier film for dried seeds packaging

A customer asked for a peelable film to package 1 kg of dried round-shaped seeds in a pillow pouch. The pouch has dimensions of 35 × 25 cm. The seeds have an initial moisture content of 5% and their maximum allowed moisture content is 10%. Determine which of the following sealants could be a good option for this pouch if one year of shelf life is needed for this product.

Table 9.1: Different sealant options for packaging round seeds products.

Sealant	Seal strength (N/25 mm)	Yield strength (MPa)	Water vapor permeability* (g μm/(m² day atm))	Price ($/lb)
B-01	3.5	10	375	$
B-02	7	10	125	$$$
B-03	5.2	11	250	$$

*At 38 °C and 90% relative humidity (RH).

9.6.1 Solution

In the case of barrier films, two main possible scenarios could happen and are shown in Figure 9.5. The first scenario (Figure 9.5(a)) is where the humidity inside of the package is higher than the outside. Packaging of moist products or freezer applications are examples of this scenario. Here, the goal for barrier packaging is to avoid moisture loss from the package. The second scenario (Figure 9.5(b)) is where the humidity inside the package is lower than the outside. In this case, the goal is to prevent moisture penetration into the package. Cereal, dried seeds and dried fruits, flours, and sugar are some examples of this scenario.

Similar to example 1 in this chapter, the required seal force for 1 kg of product can be estimated as 2.9 N/25 mm using equation (9.2). This indicates that B-01 could not be a reliable option as its average seal strength is close to the minimum seal requirement. Reformulating equation (9.3) and by using the yield strength of different film options listed in Table 9.1, the required thickness to withstand the force applied by the product weight can be estimated by

$$t = \frac{1.5 \times F}{0.025 \times \sigma} \tag{9.6}$$

This results in film thicknesses of 17.4 μm and 15.8 μm for B-02 and B-03 films, respectively.

In order to verify moisture barrier requirements, we can use the transmission rate formula [8]:

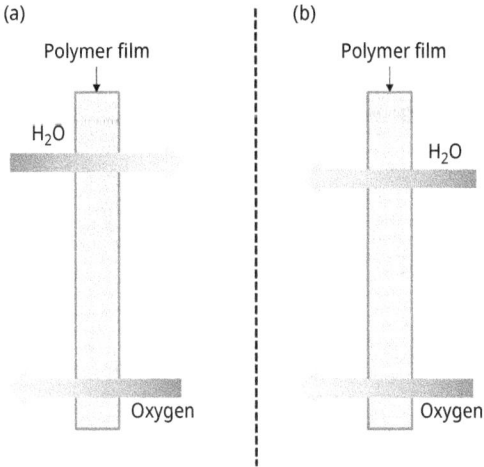

Figure 9.5: (a) Moist products where moisture or other atmospheric gases permeate out of the package and oxygen permeates toward inside the package; (b) dried products where moisture and oxygen permeate toward inside of the product. The color gradients represent schematically the concentration gradients.

$$TR = \frac{q}{t \times A \times \Delta p} \tag{9.7}$$

where TR is the transmission rate (g/(m^2 day atm)), q is the quantity of permeant, t is the permeation time, A is the package surface area, and Δp is the partial pressure difference between outside and inside of the package. As the allowed moisture intake is 5% of the product weight, the quantity of the allowed moisture intake can be calculated as 50 g. To achieve a precise design, the average temperature and humidity that the package experiences during its life cycle should be considered in permeation calculations but as the water vapor transmission rate (WVTR) measurements are commonly done by ASTM F1249 or ASTM E96 at 38 °C and 90% RH, we will use these conditions to determine the shelf life. Considering 50 mmHg or 0.0655 atm for the saturation water vapor pressure at 38 °C [9], the partial water vapor pressure at RH = 90% can be estimated as 0.059. Then the maximum allowed transmission rate for 365 days for the package with 0.175 m^2 surface area can be calculated as 13.2 g/(m^2 day atm).

Transmission rate and permeability can be related by the following equation:

$$TR = \frac{Pr}{thickness} \tag{9.8}$$

where Pr is the permeability of the film. Considering the water vapor permeability of B-02 and B-03 and the estimated film thickness, we achieve WVTR of 7.2 and 17.5 g/(m^2 day atm), respectively. These results may imply that B-03 is not a good option for this application. However, as mentioned in the beginning of this chapter, these calculations should be used as guideline and should not be solely used to design without considering technology limitations and application considerations. For example, most peelable technologies require a minimum of 5–10 μm of the peelable layer thickness to perform as expected. Considering a three-layer structure with 25% peelable layer ratio, this leads to a minimum thickness requirement of 20–40 μm. Considering an average thickness of 30 μm, both B-02 and B-03 can provide enough moisture barrier for the application. By comparing their prices, it can be concluded that B-03 is actually a better choice for this application.

9.7 Oxygen barrier sealant for cheese packaging

A customer would like to package 2.3 kg of grated parmesan cheese in a pillow pouch format with dimensions of 40 × 30 cm. The package is filled on a VFFS machine and should provide 120 days of shelf-life. Which of the following films could be the choice for this application by considering a maximum of 5 ppm allowed oxygen intake before spoilage?

	Hot tack (N/25 mm)	Yield strength (MPa)	Oxygen permeability (cc. 25 μm/(day m² atm))	Price ($/lb)
BO2-1	9	10	3.2	$
BO2-2	12	12	1.6	$$
BO2-3	10	11	1.3	$$

9.7.1 Solution

As mentioned in the previous examples where the product is packaged by VFFS, the hot tack strength should be considered as the sealant design parameter. Using equation (9.2), the required hot tack to handle 2.3 kg of the product can be estimated as 5.75 N/25 mm. Using equation (9.6), the minimum film thickness required for the three above options to handle this force can be estimated as 34.5, 28.7, and 31.3 μm for BO2-1, BO2-2, and BO2-3, respectively.

In order to examine oxygen permeation to achieve the desired shelf-life, we first need to determine the volume of oxygen gas intake. Considering the maximum allowed oxygen intake of 5 ppm and by assuming no oxygen presence inside the package initially, the volume of the allowed oxygen can be determined using the following equation:

$$q = \frac{\text{Allowed } O_2 \text{ intake(ppm)} \times 10^{-6} \times \text{product weight (g)} \times \text{gas molar volume} \left(\frac{L}{mol}\right)}{\text{Gas molecular weight} \left(\frac{g}{mol}\right)}$$

$$q = \frac{5 \times 10^{-6} \times 2{,}300 \times 22{,}400}{32} = 8.05 \text{ cc} \qquad (9.9)$$

Now by considering the package surface area of 0.24 m², the transmission rate can be calculated as follows:

$$TR = \frac{8.05 \text{ cc}}{120 \text{ days} \times 0.24 \text{ m}^2 \times 0.21 \text{ atm}} = 1.33 \frac{cc}{m^2 \text{ day atm}}$$

Using equation (9.8), the transmission rate of the available film options with the thickness determined earlier can be estimated as 2.3, 1.4, and 1.04 $\frac{cc}{m^2 \text{ day atm}}$ for BO2-1, BO2-2, and BO2-3, respectively. These results indicate that BO2-2 at ~30 µm should provide both mechanical strength and oxygen barrier properties needed for the application.

9.8 Film for frozen vegetables

You were asked to design a 50 µm film for packaging of 2.5 kg of frozen vegetable mix. The package is a pillow pouch with the dimensions of 37 × 23 cm.

9.8.1 Solution

Using the film thickness and the package dimensions, the force that is applied to the seal area can be estimated using equation (9.2):

$$F = \frac{0.072 \times 2.5 \times 9.8}{0.23} = 7.67 \frac{N}{25 \text{ mm}}$$

Then using equation (9.3), we can determine the yield strength needed from the film:

$$\sigma = \frac{1.5 \times 7.67}{0.025 \times 50 \times 10^{-6}} = 9.2 \text{ MPa}$$

It should be considered that this is the required yield strength of the film at the application temperature (−18 °C). As reducing temperature increases the yield strength of plastic materials [10], a lower yield strength at room temperature would be enough to handle the product weight. The variation of the yield strength by temperature has not been formulated due to the effects of different parameters such as chain microstructure, molecular weight, and molecular weight distribution.

Another aspect that needs to be verified for this application is the bag drop test. In this test, the product is taken out of the fridge and dropped from 3 to 5 ft height to ensure package integrity. Although designing frozen packaging might not be straightforward because of temperature, as a rule of thumb, PE-based packages for frozen food should be composed of major phases of low-density PE with the density of 0.910–0.92 g/cm^3. Therefore, the proposed film for the application in question could be any available low density film with seal force above 7.7 N/25 mm and yield strength of 9.2 MPa.

References

[1]   Ebnesajjad, S., *Plastic Films in Food Packaging: Materials, Technology and Applications*. 2012, William Andrew.

[2]   Tabatabaei, S.H., et al., *Effect of machine direction orientation conditions on properties of HDPE films*. Journal of Plastic Film & Sheeting, 2009. **25**(3–4): p. 235–249.

[3]   Clark, T.A., and J.R. Wagner Jr, *Film properties for good performance on vertical form-fill-seal packaging machines*. Journal of Plastic Film & Sheeting, 2002. **18**(3): p. 145–156.

[4]   Umezaki, E., Y. Shinoda, and K. Futase. Liquid behavior in containers with a liquid-packing bag for liquid products subjected to drop impact. In *Key Engineering Materials*. 2006, Trans Tech Publ.

[5]   Chasis, D.A., *Plastic Piping Systems*. 1988, Industrial Press Inc.

[6]   Mechtly, E., *The International System of Units: Physical Constants and Conversion Factors*. Vol. 7012. 1964, Scientific and Technical Information Division, National Aeronautics and . . . .

[7]   Kanani Aghkand, Z., and A. Ajji, *Squeeze flow in multilayer polymeric films: Effect of material characteristics and process conditions*. Journal of Applied Polymer Science, 2022. **139**(13): p. 51852.

[8]   Koros, W.J., Barrier polymers and structures: Overview. In *Barrier Polymers and Structures*, 1990, American Chemical Society. p. 1–21.

[9]   Thomson, G.W., *The Antoine equation for vapor-pressure data*. Chemical Reviews, 1946. **38**(1): p. 1–39.

[10] Landel, R.F., and L.E. Nielsen, *Mechanical Properties of Polymers and Composites*. 1993: CRC press.

Index

1-Butene 52
1-Hexane 52, 53
1-Octene 52, 53

Abuse Layer 41, 80, 81
Acid Copolymer 58
Acidic Foods 49
Adhesive Peeling 40, 43
Adjacent Reentry 23, 25
Airborne Ultrasound 37
Aliphatic Polyesters 122
Alox 1
Alpha Olefin Copolymer 52, 53, 54, 74
Aluminum 4
Aluminum Foil 59
Amorphous 15, 26, 52, 56, 71
Amorphous PET, aPET 61
Antioxidant 50
Aromatic Copolyesters 124
Atactcic PP, aPP 55, 56

Barrier 1, 49
Belt Pick-Up 7
Biaxially Oriented PP, BoPP 56
Bio-Based PE, Bio-PE 120
Bio-Based Polyethylene Terephthalate, Bio-PET 120
Biodegradable 119–122
Bioplastics 119, 120, 125
Blending 54, 58, 101, 121–124
Block Bottom 7
Bopet 61, 122
Bottom Seal 6
Breakup 105
Brittle Sealing 31
Brownian Motion 95
Bubble Test 34
Bu-LLDPE 53
Burst Test 44, 46

Cap Sealing 4
Capillary Number 105
Carreau–Yasuda Fluid Model 90, 91, 93
Cast Polypropylene, cPP 56
Caulkability 69, 85, 86, 101, 111, 134, 135
Cellophane 1

Cellulose 1
Cereal Packaging 129
Chain Diffusion 14
Chain Dynamics 18
Chain Extender 125
Chain Mobility 17
Chain Pull-Out 66
Cheese Packaging 138
Chemical Resistance 49, 58
Clarity 15, 25, 26, 39, 50, 52, 56, 121
Coalescence 105–107
Coating 1
Cohesive Peeling 40, 43, 111, 112
Compatibility 103
Compostable 119–121
Comsol 80–83, 88, 93
Consistency Index 91
Cooling Rate 26
Cooper–Mikic–Yovanovich Model, CMY model 81
Cox–Merz Law 60
Crack Healing 94
Creep Test 45
Creep To Failure 45
Critical Capillary Number 105
Critical Nucleus 24
Cross-Linking 15
Crystal Growth 23, 25
Crystal Nuclei 23
Crystal Structure 15, 23
Crystallinity 50, 52, 53, 55, 56, 69, 71, 84, 93, 121, 123, 125
Crystallization 15, 23, 25, 39, 77

Dancer Arm 6,7
Degradation 49
Delamination 41
Delay Time 40
Detergent 49
Diffusion Coefficient 19, 21, 95, 96
Disentanglement Time 20, 94
Dwell Time 5, 14, 18, 40, 63, 66–69, 77, 81, 85, 87, 96
Dye Penetration Test 36
Dynamic Scanning Calorimetry, DSC 16, 17, 25, 52, 71

https://doi.org/10.1515/9781501524592-010

Ecoflex 124
Ecovio 125
Entanglement 15, 19, 39, 50, 66, 67, 72–74
Entanglement Time 20
Enthalpy 15
Enthalpy Of Crystallization 24
Enthalpy Of Mixing 101, 102
Entropy 15
Entropy Of Mixing 101, 102
Ethylene Vinyl Acetate, EVA 1, 56–58, 63, 71,
    109, 113
Ethylene-Acrylic Acid, EAA 58, 59
Ethylene–Methacrylic Acid, EMA 58, 59, 113
Exothermic 25
Extrusion Coating 59

Failure Mode 40
Fibrillar Morphology 105, 107
Fin Seal 5,6
Final Plateau Temperature, Tpf 65
Finite Difference Method, FDM 91
Finite Element Method, FEM 82
Flat Bar Test 33
Flexural Strength 43
Flow Wrapper 7,8,9
Folded Chain Model 23
Forming Collar 6
Forming Tube 5,7
Fourier Transform Infrared, FTIR 50, 54, 57, 58, 60
Fourier'S Law 82
Fracture Energy 94, 95
Fracture Stress 94, 95
Free Volume 22, 64, 83

Gap Conductance 80
Gas Conductivity 81
Gaussian Chain 19, 20
Geometric Mean 103
Gibbs Free Energy 15
Gibbs'S Free Energy Of Mixing 101
Glass Transition Temperature, Tg 59, 69, 81, 122
Grease Resistance 61
Gross Leak 34
Gusset 8, 9, 33, 36

Harkins' Equations 107
Harmonic Mean 103
Headspace Gas 34
Heat Bar Sealing 4

Heat Conduction 80
Heat Convection 80
Heat Flux 82
Heat Of Fusion 16, 17, 71, 82
Heat Sensitive 58
Heat Transfer 14, 15, 23, 82, 85, 87, 91, 96
Heat Transfer Coefficient 80
Heavy-Duty Sack, HDSS 131, 132
Heterogeneous Nucleation 24, 25
Hex-LLDPE 53
HFFS Pouch Machine 7,8,9
High-Density Polyethylene, HDPE 50, 63, 65, 67, 69,
    71, 85, 107, 129
High-Pressure Tubular Reactor 50
Homogeneous Nucleation 23, 24, 25
Hookean Springs 19, 20
Horizontal Form Fill And Seal, HFFS 5,7,8, 39
Hot Air 5
Hot Tack 15, 22, 38, 39, 53, 58, 59, 61, 63, 66, 67,
    69, 72, 73, 93, 94, 103, 104, 122, 129, 130, 138
Hydrogen Bonding 58, 102, 122
Hydrophobic 49

Immiscible Polymer Blends 101, 102, 107, 109, 123
Impulse Sealing 9
Induction Chamber 4
Induction Sealing 4
Interdiffusion 13–15, 17, 18, 22, 51, 54, 77, 93–95
Interdigitation Time 22
Interface Healing 18
Interface Temperature 14,15, 63, 77–80, 82, 85,
    94, 96
Interfacial Tension 103, 105–109, 112
Intermittent VFFS 7
Intermolecular Interactions 15
Internal Pressurization Failure Resistance 44
Iodopovidone 36
Ionomer 59–61, 111
Isotactic PP, iPP 55, 56

Jaw Draw-Off 7

Knife 7

Lamellar Morphology 105, 107
Lap Seal 5,6
Laplace Law 106
Leakage 32, 36
Lidding 4

Linear Low-Density Polyethylene, LLDPE 51, 63–65, 69, 80–82, 102

Liquid Packaging 130

Lock Seal 41, 65, 112

Long Chain Branches, LCB 50, 72, 73, 102, 103

Low-Density Polyethylene, LDPE 50, 65, 67, 69, 102, 110

Lubricating Assumption 88

Mater-Bi 125

Mean Square Displacement 19

Mechanical Sealing 2

Mechanical Strength 50, 56, 61, 121

Melt Strength 51, 52, 60, 121

Metallocene Catalyst 54

Metallocene Polyethylene, mPE 53, 58, 69, 72. 73, 102, 103, 111

Microbial Ingress 31

Microbubbles 32

Microchannels 31

Microhardness 81

Micro-Roughness 13, 67, 80

Migration 49

Minimum Inflation Pressure 34

Misalignment 31

Miscible Polymer Blends 101, 102

Modeling 77

Moisture Barrier 1, 49, 50

Moisture Sensitive 61

Molecular Architecture 49, 52, 54, 72

Molecular Mobility 14, 15, 25, 54

Molecular Weight Distribution, MMD 16, 17, 21, 53, 72, 93, 139

Molecular Weight, Mw 16, 17, 19, 20, 21, 56, 72–74, 93, 102, 139

Monoaxially Oriented PP, MoPP 56

Morphology 60, 63, 101, 104, 107, 111, 122

Multicomponent Sealant 101

Multiwall Paper Bags, MWPB 3, 50

Nanocomposite 112

Narrow Seals 31

Newtonian Fluid 86, 87, 93

Niche Distance 25

Nichrome 9

Nitrocellulose 1

Nonisothermal Crystallization 23

Non-Newtonian Fluid 86

Nucleating Agent 56

Nucleation 23

Oct-LLDPE 53

Optimization 77

Organically Modified Montmorillonite, oMMT 112

Orientation 17, 56, 61, 129

Oversealed Area 31

Over-The-Mountain, OTM 132, 133

Oxygen Barrier 50, 139

Partially Miscible Polymer Blends 101, 125

Peelable 41, 63, 110–112, 129, 132, 133, 136

PE-g-MA 83, 84, 112

Permeation 49, 137

Pharmaceutical 61

Pinholes 32

Plastomers 54, 71, 111

Plateau Initiation Temperature, Tpi 65, 71, 96

Plateau Seal Strength 65

Poly(3-Hydroxybutyrate-Co-3- Hydroxyvalerate), PHBV 125

Poly(Butylene Adipate-Co-Terephthalate), PBAT 107, 119, 124

Poly(Butylene Succinate/Adipate), PBSA 122

Poly(Ethylene Terephthalate), PET 1, 17, 61, 120

Poly(Hydroxybutyrate), PHB 125

Poly(Lactic Acid), PLA 107, 121–125

Polyamide, PA 1, 81, 83

Polybutene-1, PB-1 110

Polybutylene Succinate, PBS 122

Polycaprolactone, PCL 122

Polyethylene, PE 1, 17, 49, 109, 110, 119

Polyhydroxyalkanoates, PHA 125

Polymer Blends 101

Polymer Hardness 14

Polypropylene, PP 1, 17, 54, 56, 69, 71, 109, 125

Polystyrene, PS 17, 107

Polyvinyl Chloride, PVC 1

Polyvinylidene Chloride, PVDC 1

Power-Law Index 91

Power-Law Model 90

Pressure Decay Leak Test 35

Pressure-Assisted Dye Penetration Test 37

Primary Crystals 23

Printability 61

Radical Polymerization 50, 56

Radius Of Gyration 22, 96

Random Copolymerization 51, 124
Recrystallization 13
Relaxation Time 19, 20
Reptation Model 20, 21, 67, 94
Rigid Containers 4
Rotating Heated Seal Bar 9
Rouse Model 18, 19, 20, 21
Rouse Relaxation Time 20

Scaling Law 94
Seal Initiation 53
Seal Initiation Temperature, Tsi, SIT 65, 67, 71–74, 93, 102, 107, 109, 111, 122, 125
Seal Performance Test 31, 38
Seal Quality Tests 31
Seal Strength 23, 39, 42, 56, 63, 93–96, 102, 109, 122, 129, 132
Seal Through 61
Sealant Materials 49
Sealing Pressure 5, 14, 63, 67, 82, 85, 92
Sealing Temperature 5, 63, 66, 67, 69
Sealing Window 65, 101
Serrated Jaws 5
Shear Rate 105
Short Chain Branches, SCB 51, 72, 73
Side-Chain Groups 51, 53
Simulation 77, 81, 88
SiOx 1
Sonotrode 3
Specific Heat 79, 82, 83, 85
Specific Volume 16
Spherulites 56
Spouted Pouch 134, 135
Spreading Coefficients 108
Springback Force 39
Squeeze Out Flow, SOF 65, 67, 69, 77, 85–93, 135
Statistical Segment 19
Steric Hindrance 56
Sterilization 31
Subcritical Nuclei 24
Supercooling 23, 25
Supercritical Nuclei 25
Surface Asperities 15, 17,50, 81, 87
Surface Impurities 18
Surface Rearrangement 15, 17, 18
Surface Roughness 17, 81

Surface Tension 103
Surface Wetting 17, 67, 94
Switchboard Model 23, 25
Syndiotactic PP, sPP 55, 56

Tacticity 55, 56
Temperature Gradient 78, 82
Temperature-Modulated DSC, TMDSC 52, 84
Temperature-Sensitive Materials 3
Tensile Machine 44
Terminal Relaxation Time 96
Thermal Conductivity 15, 79, 82–84
Thermal Contact Resistance, TCR 80, 81
Thermal Contact Theory 80
Thermal Diffusivity 87
Thermal Oxidation 54
Thermal Stability 49
Thermodynamics 15, 23, 101, 107, 108, 112
Thermoplastic Starch, TPS 122, 123, 125
Transport Belts 6
Twin-Jaw Continuous Motion 7

Ultrasonic Anvil 2,3
Ultrasonic Horn 2,3
Ultrasonic Sealing 2, 50
Unsealed Area 31

Vacuum Decay Leak Test 36
Vacuum Packing 34
Vacuum Transport Belts 7
Vertical Form Fill And Seal, VFFS 5,6,7, 39, 129, 130, 138
Vertical Sealer 5
Very Low Density Polyethylene (vLDPE) 69, 109
Viscosity 19, 20, 87, 88, 105, 108, 111, 113
Visual Inspection 33

Water Vapor Transmission, WVTR 137
Welding 2, 22
Work Of Adhesion 103

Young'S Model 112, 113

Zero Shear Viscosity 21, 85
Ziegler–Natta 53, 54
Zipper 9

www.ingramcontent.com/pod-product-compliance
Lightning Source LLC
Chambersburg PA
CBHW081539220326
41598CB00036B/6486